MINITAB Guide to The Statistical Analysis of Data

T. W. Anderson, Stanford University

Barrett P. Eynon, SRI International

▲ *The Scientific Press*
540 University Avenue
Palo Alto, CA 94301
(415) 322-5221

MINITAB GUIDE TO THE STATISTICAL ANALYSIS OF DATA
T. W. Anderson and Barrett P. Eynon

ISBN 0-89426-072-3

Cover Design by Rogondino & Associates

PREFACE

This *Guide* is a supplement to *The Statistical Analysis of Data,* Second Edition, by T. W. Anderson and Stanley L. Sclove, providing an introduction to computational statistics with the Minitab computer package. Hands-on data analysis experience with the Minitab package provides benefits both in the learning and the application of the statistical concepts described in the text.

This *Guide* is organized along the lines of the text. Every procedure presented in the text (with very few exceptions) can be carried out with Minitab; each is fully explained in this *Guide*. All information necessary for the use of Minitab has been included. A command reference is given at the end of each chapter, and a complete list of Minitab commands is included as an appendix. The *Guide* and the text, taken together, form a self-contained instructional package.

The computer exercises given in the *Guide* have been chosen to illustrate the concepts developed in each chapter and their applications to real datasets. In order to facilitate the working of the exercises, a number of the example data sets used in this *Guide* are available from the publisher on an IBM PC diskette and also will be included with future Minitab distributions for mainframe use.

The Minitab examples presented in this *Guide* were performed with the 1985 version of Minitab. Differences from the 1982 version of Minitab are noted for the benefit of those using the earlier version.

The authors would like to thank Barbara Ryan of Minitab, Inc., for providing the current version of Minitab and for her comments and suggestions, and also Judi Davis for assisting with the data entry. This *Guide* was prepared using the LaTeX macro package for the TeX typesetting system and reproduced by The Scientific Press from camera-ready copy supplied by the authors.

T. W. A.
B. P. E.

CONTENTS

**PART FIVE   STATISTICAL METHODS
                FOR OTHER PROBLEMS   133**

PART ONE

INTRODUCTION

1 GETTING STARTED WITH MINITAB

INTRODUCTION

Minitab is an easy-to-use interactive computer program for statistical analysis. Use of such a computer package eliminates much of the hand calculation of statistics and allows the analysis of much larger data sets. Since it is interactive, different analyses can be tried out, and the results appear immediately.

Minitab is available for a number of different computer systems, including both mainframe and personal computers.

1.1 STARTING MINITAB

The details of starting the Minitab program will vary depending on the computer system you are working with. If you are working on a multiuser computer, you will probably first have to identify yourself to the computer with your user name or account and password. If you are working on a personal computer, you will need to turn the computer on and insert the Minitab program disks.

Following this you will probably give the command "MINITAB" to your computer in order to start Minitab running. See your Minitab program documentation or system documentation for details.

Once you give this command, Minitab will respond with a greeting message and then prompt you with a MTB > prompt symbol to let you know it is awaiting your commands. For instance

```
@MINITAB
MINITAB RELEASE 5.1 *** COPYRIGHT - MINITAB, INC. 1985
U.S. FEDERAL GOVERNMENT USERS SEE HELP FGU
DEC. 9, 1985 *** Stanford University
DEC-20/TOPS-20, Storage available 100000
 Welcome to Minitab, version 5.1 for TOPS-20.
MTB >
```

You can now type commands to Minitab, and Minitab will print the results from that command followed by another MTB > prompt for your next command. For instance, you can calculate the sum of a set of numbers as follows:

```
MTB > SET DATA INTO C1
DATA> 1 4 2 6 5
DATA> END OF DATA
MTB > SUM OF C1
    SUM    =      18.000
```

When you are done with your Minitab session, type the command

STOP

to the MTB > prompt to leave the program.

```
MTB > STOP
*** Minitab Release 5.1 *** Minitab, Inc. ***
DEC-20/TOPS-20, Storage available 100000

CPU time 3.77   Elapsed time 14.85
@
```

1.2   THE MINITAB WORKSHEET

Think of Minitab as having a large sheet of lined paper with many vertical columns and horizontal rows on it. Each column is labeled at the top with C1, C2, C3, etc. (up to C50, or more in some versions of Minitab). Each column can contain a set of numbers with which you are working, each number in its own row. The numbers in a column will all be of a type. For instance, if we were using Minitab to store course grades, we might put midterm scores into column 1 (C1), and final scores into column 2 (C2). In this example, each row would correspond to a particular student, much as with a physical grade book. At your command Minitab will write numbers into a column, calculate statistics on the numbers in the column, add two columns together row-by-row, or print out the contents of a column. In addition to the columns of the worksheet there are stored constants named K1, K2, K3, etc., that can each contain a single number, and matrices named M1, M2, M3, etc., that can each contain a two-dimensional array of numbers (a two-way table). (We shall not discuss the use of matrices in this *Guide*.)

1.3   MINITAB COMMANDS

Commands are typed by the user at the MTB > prompt, and consist of a command name followed by the items to be operated on. Commands can be typed in either upper or lower case.[1] Each command must start on its own line. At the end of the command line, you type a carriage return (usually marked CR or Return on your keyboard) to tell Minitab to perform the command. Until you type the carriage return, you can use the Backspace key to back up and retype parts of the command.

In this *Guide* commands are presented in the format

COMMAND operating on arguments **C**, **K**, **M**, etc.

where **C**, **K**, **M**, etc. stand for the types of arguments that may be used with that command. Each time **C** is mentioned, we put in the appropriate column name. For instance, the ADD command is defined as

ADD C to **C**, put result in **C**

We substitute for **C**'s the designation for the particular columns we want to operate on and obtain, for instance,

```
MTB > ADD C1 TO C2, PUT RESULT IN C3
```

This adds the contents of column 1 (C1) to that of column 2 (C2) row by row, and puts the result in column 3 (C3). A **K** in a command description can be either a stored constant (K1, K2, etc.) or a number (3 or 27.5). An **E** in a command description refers to either a column or a constant.

Some commands can optionally operate on several columns at once, or both print their results and and store the answer in a column or a stored constant for further analysis. Optional parts of commands are indicated in square brackets ([ ]), and the arguments inside the brackets can be omitted if the optional results are not needed. For instance, the PRINT command can be used to print the contents of one column or several columns. The square brackets are not actually typed as part of the command.

Lists of columns can be indicated by a dash, e.g.,

```
MTB > PRINT C1-C3
```

and

```
MTB > PRINT C1 C2 C3
```

are equivalent.

Command names can be abbreviated to their first four letters. Extra text in addition to the first four letters of the command name and references to columns (C), constants (K), matrices (M), and actual numbers is ignored by Minitab but is helpful to you for remembering what the computations are. The extra words in the command descriptions in this *Guide* are for describing the use of the command and identifying the arguments, but do not have to be typed as part of the command. The positions of the arguments in a command line determine how they are used. So instead of

```
MTB > ADD C1 TO C2, PUT RESULT IN C3
```

you could type

```
MTB > ADD C1 C2 C3
```

It is important to keep in mind that Minitab does not read any extra text on the command line. If you type

```
MTB > DIVIDE C1 INTO C2 AND PUT THE RESULTS INTO C3
```

you will not have what you expect in C3, since this is read by Minitab as

```
MTB > DIVI C1 C2 C3
```

and the command (as we shall see below) is actually

DIVIDE C *by* C, put result in C

Examples of actual Minitab use are given throughout this *Guide*. Lines preceded by MTB > are the lines you type to Minitab following the prompt. (The MTB > is Minitab's signal that it awaiting your input.) All of the other lines are what Minitab types back.

If you happen to mistype a command name or give the wrong number or type of arguments to a command, Minitab will print an error message:

```
MTB > PRIMT C1 AND C2
* ERROR * NAME NOT FOUND IN DICTIONARY
        * MISSPELLED NAME, OR ERROR IN READ OR SET, OR DATA OUT OF PLACE

MTB >
```

If this happens, you can simply retype the correct command to the next MTB > prompt, and continue with your session.

If you need to type a command that is longer than will fit on one line, simply end the line with an & and continue on the next line.[2]

1.4   ENTERING DATA INTO MINITAB

To enter data into Minitab for computation, we use the command

READ the following data into columns C,...,C

Following the command we type the data for each column, one line per row. For instance if we want to set up the worksheet with the information in Table 1.9. These might be the scores on two midterm examinations for four students in a course. Each column corresponds to an examination, and each row corresponds to a student. To read the data we do as follows:

TABLE 1.9
A Simple Data Set

Column 1	Column 2
100	36.0
105	37.3
99	42.6
115	40.0

```
MTB > READ DATA INTO C1 AND C2
DATA> 100 36
DATA> 105 37.3
DATA> 99   42.6
DATA> 115 40
DATA> END OF DATA
      4 ROWS READ
```

During data entry, Minitab displays a DATA> prompt. Numbers are separated by spaces or commas. Any other text on the data lines is ignored, including extra spaces. Numbers can be typed with or without a decimal point. Commas should not be used in large numbers, as Minitab will think two numbers are being typed. Each line must have just enough entries for all of the columns in the READ command. (Minitab will ask you to retype the line if too few or too many numbers are typed.) Minitab will continue reading data until the next command line is entered; alternatively the optional command

END of data

can be used to indicate the end of the data. (The END command is never necessary, but sometimes results in tidier output, as the partial data listing Minitab produces after reading large data sets will be properly sequenced in the output.)

If we need to enter only one column of numbers, it may be more convenient to type all of the numbers across the line. The command

SET the following data into column **C**

will put all numbers on following lines into the specified column. Several lines of numbers may be entered. For example

```
MTB > SET DATA INTO C3
DATA> 83 47 52.4
DATA> 94.2
DATA> END OF DATA
```

puts all four numbers into C3.

1.5   PRINTING DATA FROM MINITAB

To print data from Minitab we use the command

PRINT C (or **K** or **M**), ... , **C** (or **K** or **M**)

If only one column is printed it is printed across the page to save space; otherwise columns are printed vertically down the page. For our example data above we have

```
MTB > PRINT C1,C2, AND C3
 ROW     C1     C2     C3

   1    100    36.0   83.0
   2    105    37.3   47.0
   3     99    42.6   52.4
   4    115    40.0   94.2

MTB > PRINT C3
C3
  83.0    47.0    52.4    94.2
```

1.6   TRANSFORMING DATA INSIDE MINITAB

1.6.1   Arithmetic Operations

Once our data are in Minitab there are many commands to calculate new columns from the existing ones. The following commands perform the usual arithmetic operations row by row on columns (or constants).

ADD E to **E**, put into **E**

For each **E** in the command description we can use either a column (such as C1), a stored constant (such as K1) or a number (such as 5). Adding a constant to a column adds the constant to each row. (This holds for all of the functions in this section.) For example:

```
MTB > ADD C1 TO C2, PUT IN C4
MTB > ADD 5 TO C4, PUT IN C5
MTB > PRINT C1 C2 C4 C5
 ROW    C1     C2      C4      C5

   1   100    36.0   136.0   141.0
   2   105    37.3   142.3   147.3
   3    99    42.6   141.6   146.6
   4   115    40.0   155.0   160.0
```

SUBTRACT E from E, put into E

MULTIPLY E by E, put into E

DIVIDE E by E, put into E

RAISE E to the power E, put into E

If the result of a command is undefined, such as the result of division by 0, the entry is set to *, which is the Minitab indicator of missing or undefined values.

1.6.2 Functions

The following functions operate on either a column or a constant.

ABSOLUTE value of E, put into E

This calculates the absolute value of each number, i.e. both -3 and 3 have an absolute value of 3.

SIGNS of E, put into E

This calculates the "sign" of each number, giving -1 for negative numbers, +1 for positive numbers, and 0 for 0. A summary count of the signs is printed.

SQRT of E, put into E

This calculates the square root of each number. The square root of a negative number is undefined, and Minitab will set the result to the * missing value indicator in such cases.

Other commands to calculate mathematical functions such as logarithms and trigonometric functions are also available in Minitab.

1.6.3 The LET command

A command that allows more complicated expressions is

LET E = expression

For example

```
MTB > LET C6 = (C1 + C2) / C3
```

The LET command allows us to perform calculations that would take a series of commands otherwise. For the above expression we would have to do

```
MTB > ADD C1 TO C2, PUT IN C7
MTB > DIVIDE C7 BY C3, PUT IN C6
```

to achieve the same result. The expressions in a LET command can use +, −, *, and /, to mean plus, minus, times, and divide by, respectively. Exponentiation (raising to a power) is indicated by **. Parentheses control the order of evaluation as per algebraic conventions.[3] For instance both

```
MTB > LET C6 = C1 + C2 / C3
```

and

```
MTB > LET C6 = C1 + ( C2 / C3 )
```

first result in the entries in C2 being divided by those of C3, and the result added to C1, whereas in

```
MTB > LET C6 = (C1 + C2) / C3
```

the entries in C1 and C2 are summed, and the result divided by C3.

LET expressions can include columns, numbers, and stored constants. Many functions and column operations corresponding to Minitab commands are also available in LET expressions, including ABSOLUTE, SIGNS, and SQRT, and most of the statistical functions we will learn about later. These functions are used by putting parentheses around the argument, e.g.

```
MTB > LET C2 = SQRT(C1)
```

No extra text is allowed in a LET expression.

Individual rows of a column can be referred to by putting the row number in parentheses after the column, e.g., C3(4) referrs to row 4 of column 3.

1.7 DOCUMENTING YOUR WORK

It is a good idea to put comments in your work, so that when you look back at the output or when someone else reads it there will be a record of what you were doing. Using extra words in command lines is a good way to document each command. In addition the command

NOTE - comments go here

can be used to put in a line of comments.

It is also very useful to give names to the columns so you know what they contain. The command

NAME for C is 'NAME' ..., for C is 'NAME'

gives names to columns. Each name may be up to 8 characters long. Anywhere a column reference is allowed, the name (in single quotation marks) can be used to refer to the column. Minitab will also use the name to label its output. Names must always be referred to in quotes, otherwise Minitab will ignore them as extra text.

Here is an example of all three kinds of documentation:

```
MTB > NOTE HERE IS AN EXAMPLE OF NAMING COLUMNS
MTB > NAME FOR C1 IS 'SCORES'
MTB > PRINT 'SCORES'
SCORES
    100    105    99    115
```

You can also put comments at the end of a command line by typing a # followed by your comment. Minitab ignores everything following the #.[4]

1.8 GETTING HELP

The command

HELP about **TOPIC**

can can be used to obtain help about any Minitab command. Saying HELP with no topic will print out a general introduction to the HELP facility. Saying HELP OVERVIEW will get general help. Saying HELP COMMANDS will give a list of all HELP topics available.

COMMAND REFERENCE

Listed below are the general forms of the Minitab commands discussed in this chapter, with additional notes. A complete listing of Minitab commands is given in the appendix.

READ the following data into columns **C**,...,**C**
END of data
SET the following data into column **C**
ADD **E** to **E**, [to **E** to **E** ...] put into **E**
    Several columns may be added at once using the extended form.
SUBTRACT **E** from **E**, put into **E**
MULTIPLY **E** by **E**, [by **E** by **E** ...] put into **E**
    Several columns may be multiplied at once using the extended form.
DIVIDE **E** by **E**, put into **E**
RAISE **E** to the power **E**, put into **E**
ABSOLUTE value of **E**, put into **E**
SIGNS of **E**, put into **E**
SQRT of **E**, put into **E**
LET **E** = expression
PRINT (**C** or **K** or **M**),...,(**C** or **K** or **M**)
NOTE - comments go here
NAME for **C** is 'NAME', ..., for **C** is 'NAME'
HELP about **TOPIC**
STOP - exit Minitab

COMPUTER EXERCISES

1.1 Log on to your computer account or turn on your microcomputer and set it up to use Minitab. Start up Minitab and try out a few of the commands in this chapter. In particular, READ some numbers into a column, and PRINT them out. Try the HELP facility. Try anything else that interests you. Then give the STOP command to leave Minitab, and log off your computer.

NOTES

1. In some versions of Minitab, commands must be typed in upper case letters (that is, capitals) only. (You can depress the Caps Lock key on your computer or terminal while using Minitab.)

2. In pre-1982 versions of Minitab, the continuation command was *.

3. The default order of precedence for evaluating LET expressions is first functions and subscripts (i.e. references to specific rows of a column), then **, then * and /, and then + and −. Operators with the same precedence are evaluated from left to right.

4. Not available in earlier versions of Minitab.

PART TWO

DESCRIPTIVE STATISTICS

2 ORGANIZATION OF DATA

INTRODUCTION

A computer program such as Minitab provides an easy and convenient way to work with statistical information. Once the data are set up for use in the computer, the information can easily be organized and displayed, in many useful ways, including tables, graphs, and charts.

We saw in Chapter 1 how to read numbers into Minitab using the READ and SET commands. In this chapter we shall see how to set up various kinds of data for use with Minitab and how to work with them to explore what the data can tell us.

2.1 KINDS OF VARIABLES; SCALES

2.1.1 Categorical Variables

Minitab can only store numbers and not words in its columns. If we have categorical information, we shall represent it in Minitab with numerical *codes* that correspond to the categories. For instance in working with the occupational data from Table 2.1, we might assign the value 1 to represent professionals, 2 to represent sales, 3 to represent clerical, and 4 to represent laborers. Then we read the data into Minitab as follows:

```
MTB > NOTE OCCUPATIONS OF HEADS OF HOUSEHOLDS
MTB > NOTE FROM TABLE 2.1 OF TEXT
MTB > NOTE OCCUPATIONS: 1=PROFESSIONAL 2=SALES 3=CLERICAL 4=LABORER
MTB > NOTE FIRST READ THE DATA INTO COLUMN 1
MTB > SET C1
DATA> 3 1 1 3 1 1 2 4 2 2 2 2 2 1 3 2 3 2 4
DATA> END OF DATA
MTB > NOTE NOW GIVE THE COLUMN A NAME
MTB > NAME C1 'OCCUP'
MTB > NOTE PRINT OUT WHAT WE HAVE ENTERED
MTB > PRINT 'OCCUP'
OCCUP
    3    1    1    1    3    1    1    2    4    2    2    2    2    2
    1    3    2    3    2    4
```

It is usually convenient to represent the categories with simple integers. If we are working with several categorical variables in a study, we shall need to set up a list of numbers to represent the categories for each variable. This is often referred to as the *coding list* for the variables.

Suppose we have made a list of friends and noted the sex, hair color, and student status of each, as shown in Table 2.29.

TABLE 2.29
Some Data on Friends

Name	Sex	Hair Color	Student?
Fred	Male	Brunette	Yes
Barbara	Female	Brunette	No
George	Male	Blond	Yes
Allen	Male	Redhead	No
Karen	Female	Blond	Yes

Then we might code hair color as blond = 1, brunette = 2, and redhead = 3. For a dichotomous variable, with only two levels, such as sex, we might choose male = 1, female = 2 (or the reverse, as long as we keep track of our coding list). For a yes-no question (whether someone is a student, whether a plant bears flowers) it is sometimes useful to code no = 0 and yes = 1. If we use the codings suggested above, we obtain the data set shown in Table 2.30. We shall not actually put the names into

TABLE 2.30
Data on Friends, after Coding

Name	Sex	Hair Color	Student?
Fred	1	2	1
Barbara	2	2	0
George	1	1	1
Allen	1	3	0
Karen	2	1	1

Minitab, but it is convenient to keep them while we are coding to make it easier to check our coding. If we want to put this data into Minitab, we do the following:

```
MTB > READ C1 C2 C3
DATA> 1 2 1
DATA> 2 2 0
DATA> 1 1 1
```

```
DATA> 1 3 0
DATA> 2 1 1
DATA> END OF DATA
      5 ROWS READ
MTB > NAME C1 'SEX' C2 'HAIR' C3 'STUDENT'
```

We can now use Minitab to work with and display this data as described in the remainder of this chapter and the rest of this *Guide*.

2.1.2 Numerical Variables

Both discrete and continuous numerical variables can be easily used in Minitab. Usually the numbers can be typed directly into Minitab. When working with data that are not integers, fractions will have to be converted to decimal values. For instance, if we were working with stock prices, a stock price of $17\frac{3}{8}$ is coded as 17.125. A convenient choice of units is often useful and makes entering data easier, for instance if we were entering housing prices, we might put housing prices in thousands of dollars, instead of dollars. If large numbers are used, do not type commas in the value; Minitab will think you are typing in several smaller numbers.

2.1.3 Scales

All of the types of scales described in the text will be represented in Minitab by numbers. The program does not know anything about the type of data in a column, so the only distinction is in how you use them and interpret the results. For categorical data it important to remember that the numbers are just codes, and that blonde plus brunette does not equal redhead!

2.1.4 Data and Observational Units

Usually each row (horizontal line) of the Minitab worksheet will represent one observational unit. The entries in each column (vertical line) corresponding to that row are the data on the variable corresponding to that column.

2.2 ORGANIZATION OF CATEGORICAL DATA

2.2.1 Frequencies

Once we have our data in Minitab there are several ways to summarize the data.

Frequency Distributions. The TABLE command is very useful, and we shall call upon it in several different ways in this *Guide*. In its simplest form it will give us the frequency counts for the values in a column:

TABLE data in C

The values in the column must be integer values.[1]

Here is how we use Minitab to analyze the occupational data:

```
MTB > NOTE CALCULATE THE FREQUENCY DISTRIBUTION OF OCCUPATIONS
MTB > TABLE FOR 'OCCUP'

  ROWS: OCCUP

        COUNT

    1      6
    2      8
    3      4
    4      2
  ALL     20
```

This gives us the distribution of occupations. To find the relative frequencies of occupations, we use a more extended form of the TABLE command, by typing a semicolon (;) at the end of the first line of the command, and then the subcommand

---
TABLE

    TOTPERCENTS - percentages of the total table
---

and a period to end the subcommand mode:

```
MTB > NOTE CALCULATE THE RELATIVE FREQUENCIES OF OCCUPATIONS
MTB > TABLE FOR 'OCCUP';
SUBC> TOTPERCENTS.

  ROWS: OCCUP

      % OF TBL

    1    30.00
    2    40.00
    3    20.00
    4    10.00
  ALL  100.00
```

Note that Minitab shifts to a SUBC > prompt while we are entering subcommands.

We can combine both types of information in the same table with the COUNTS subcommand

> *TABLE*
>
> ## COUNTS - frequency counts for each entry

plus the TOTPERCENTS subcommand:

```
MTB > NOTE PUT BOTH THE FREQS AND THE RELATIVE FREQS ON THE SAME TABLE
MTB > TABLE FOR 'OCCUP';
SUBC> COUNTS;
SUBC> TOTPERCENTS.

  ROWS: OCCUP

         COUNT % OF TBL

    1       6     30.00
    2       8     40.00
    3       4     20.00
    4       2     10.00
  ALL      20    100.00
```

Another command that will produce frequency counts[2] is

> **TALLY** data in C

with corresponding subcommands

> *TALLY*
>
> ## COUNTS - frequencies

and

> *TALLY*
>
> ## PERCENTS - relative frequencies

In the applications discussed here, the two commands produce essentially the same output. We shall discuss the different uses of the TALLY and TABLE commands in later sections.

Comparing Distributions. We can use the TABLE command to print frequencies for side by side comparison. In order to demonstrate, we shall use the example from Table 2.4 on employment categories in 1950 and 1970.

Occasionally, as in this example, when we receive data it will have already have been summarized, and the entries in each row will be the summary results over several observational units. In such cases we shall usually create an extra variable that contains the number of individuals or *frequency* in that summary group. If we were not to do this in our example, we would have to type 38,068 lines of data just for the white-collar workers in 1970! A separate variable is set up for the frequencies in each year. The new subcommand to the TABLE command we use here is

---

TABLE

FREQUENCIES in C   .

```
MTB > NOTE EMPLOYED PERSONS BY MAJOR OCCUPATION GROUPS IN 1950 AND 1970
MTB > NOTE C1: OCCUPATIONS: 1=WHITE-COLLAR 2=BLUE-COLLAR 3=SERVICE 4=FARM
MTB > NOTE C2: NUMBERS OF PERSONS (THOUSANDS) IN 1970
MTB > NOTE C3: NUMBERS OF PERSONS (THOUSANDS) IN 1950
MTB > READ C1 C2 C3
DATA> 1 38068  22373
DATA> 1 38068  22373
DATA> 2 27452  23336
DATA> 3  9724   6535
DATA> 4  3164   7408
DATA> END OF DATA
        4 ROWS READ
MTB > NAME C1 'OCCUP' C2 '1970' C3 '1950'
MTB > NOTE PERCENTAGE OF EMPLOYED PERSONS IN MAJOR OCCUPATION GROUPS
MTB > NOTE IN 1950 AND 1970
MTB > TABLE FOR 'OCCUP';
SUBC> FREQUENCIES IN '1970';
SUBC> COUNTS;
SUBC> TOTPERCENTS.

  ROWS: OCCUP

         COUNT % OF TBL

     1   38068    48.55
     2   27452    35.01
     3    9724    12.40
     4    3164     4.04
   ALL   78408   100.00

MTB > TABLE FOR 'OCCUP';
SUBC> FREQUENCIES IN '1950';
SUBC> COUNTS;
SUBC> TOTPERCENTS.

  ROWS: OCCUP

         COUNT % OF TBL

     1   22373    37.51
     2   23336    39.12
```

```
    3      6535     10.96
    4      7408     12.42
  ALL     59652    100.00
```

We can cut the these two tables from our output and place them side by side to compare the distributions of employment in 1950 and 1970.

2.2.2 Frequency Graphs

The HISTOGRAM command will give a frequency bar graph for categorical data:

HISTOGRAM for data in C

Returning to our example on occupational categories of heads of households we have

```
MTB > NOTE FREQUENCY GRAPH OF OCCUPATIONS
MTB > HISTOGRAM OF 'OCCUP'

Histogram of OCCUP    N = 20

Midpoint   Count
       1       6   ******
       2       8   ********
       3       4   ****
       4       2   **
```

Other graphical presentations can be aided by preparing the data with Minitab and then transfering the information to other computer graphics programs or preparing the charts by hand.

2.3   ORGANIZATION OF NUMERICAL DATA

2.3.1   Discrete Data

Frequency tabulations and bar charts for discrete data can be obtained in exactly the same fashion as for the categorical data above, if the data values are integers. For instance, using the data on number of children in households from Table 2.6 of the text we have

```
MTB > NOTE NUMBER OF CHILDREN IN 20 HOUSEHOLDS - TABLE 2.6
MTB > SET C1
DATA> 2 0 1 0 0 1 1 1 4 1 3 2 2 1 1 2 0 3 1 4
DATA> END OF DATA
MTB > NAME C1 'NCHILD'
MTB > TABLE FOR 'NCHILD';
SUBC> COUNTS;
SUBC> TOTPERCENTS.
```

```
ROWS: NCHILD

        COUNT % OF TBL

   0       4    20.00
   1       8    40.00
   2       4    20.00
   3       2    10.00
   4       2    10.00
 ALL      20   100.00

MTB > HISTOGRAM FOR 'NCHILD'

Histogram of NCHILD    N = 20

Midpoint    Count
       0       4    ****
       1       8    ********
       2       4    ****
       3       2    **
       4       2    **
```

Cumulative Frequencies. Cumulative frequencies and cumulative relative frequencies can be obtained with the TALLY command and the subcommands

TALLY
CUMCNTS - cumulative counts

and

TALLY
CUMPCTS - cumulative percentages

For instance, in order to obtain the cumulative frequencies on the number of children in households, we do

```
MTB > TALLY FOR 'NCHILD';
SUBC> CUMCNTS;
SUBC> CUMPCTS.

NCHILD   CUMCNT   CUMPCT
     0        4    20.00
     1       12    60.00
     2       16    80.00
     3       18    90.00
     4       20   100.00
```

Cumulative relative frequency plots can be prepared using the PLOT command

PLOT data in C versus data in C

by plotting the ordered data $X_{(i)}$ versus i/n. The data, for instance the data in Table 2.6, can be ordered by the SORT command:

SORT data in C, put in C

The value of n can be obtained with the COUNT command

COUNT the number of values in C, put result in **K**

and the plotting values i/n can be obtained by first making a column containing $1, \ldots, n$ with the GENERATE command

GENERATE integers from 1 to **K**, put in C

and then dividing the column by n.

For the example above, we have

```
MTB > NOTE CUMULATIVE RELATIVE FREQUENCY PLOT
MTB > SORT 'NCHILD', PUT C2
MTB > NAME C2 'NCHILD-S'
MTB > COUNT 'NCHILD', PUT K1
   COUNT   =      20.000
MTB > GENERATE INTEGERS FROM 1 TO K1 , PUT IN C3
MTB > DIVIDE C3 BY K1, PUT IN C4
MTB > NAME C4 'RLFRQ'
MTB > PLOT 'RLFRQ' BY 'NCHILD-S'
```

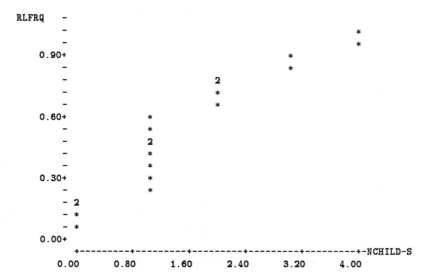

```
RLFRQ    -
         -                                                *
         -                                                *
   0.90+                                        *
         -                                        *
         -                           2
         -                           *
         -                           *
   0.60+            *
         -            *
         -            2
         -            *
         -            *
   0.30+            *
         -            *
       - 2
       - *
       - *
   0.00+
         +---------+---------+---------+---------+---------+-NCHILD-S
        0.00      0.80      1.60      2.40      3.20      4.00
```

The student should draw horizontal lines to the right from the top of each vertical line of points to complete the cumulative distribution plot.

Comparing Two Frequency Distributions. We can use the MPLOT command,

MPLOT C vs C, C vs C

which plots several pairs of variables on the same graph to compare two frequency distributions. For instance, with the data from Table 2.8 on making errors under pressure:

```
MTB > NOTE DISTRIBUTION OF ERRORS MADE BY LAST PERSON - TABLE 2.8
MTB > READ C1 C2 C3
DATA> 0 13 35          DATA> 7 2 0
DATA> 1 4 1        /   DATA> 8 5 0
DATA> 2 5 1        /   DATA> 9 3 0
DATA> 3 6 0        /   DATA> 10 3 0
DATA> 4 3 0           DATA> 11 1 0
DATA> 5 4 0           DATA> 12 0 0
DATA> 6 1 0           DATA> END
     13 ROWS READ
MTB > NAME C1 'NERR' C2 'PRESS' C3 'NOPRESS'
MTB > NOTE COMPARISON OF DISTRIBUTIONS OF NUMBER OF ERRORS
MTB > MPLOT 'PRESS' VS 'NERR' AND 'NOPRESS' VS 'NERR'
```

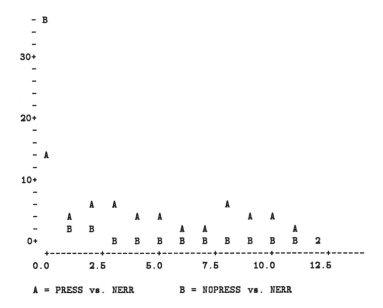

```
     - B
     -
     -
     -
  30+
     -
     -
     -
     -
  20+
     -
     -
     - A
     -
  10+
     -
     -       A   A                  A
     -    A          A   A              A   A
     -    B   B              A   A           A
   0+          B   B   B   B   B   B   B   B   B   2
     +---------+---------+---------+---------+---------+------
    0.0       2.5       5.0       7.5      10.0      12.5

   A = PRESS vs. NERR        B = NOPRESS vs. NERR
```

Here an 'A' marks the first pair of plotting variables (the data for the pressure group), and a 'B' marks the second pair of plotting variables (the data for the control group). The '2' on the plot indicates that the two values were almost in the same place and there was not room for two letters. So this plot corresponds to Figure 2.8 with 'A' instead of 'X' and 'B' instead of 'C'.

We can also compare relative frequencies if we compute them first:

```
MTB > NOTE COMPARISON OF RELATIVE FREQUENCIES OF ERRORS
MTB > LET C4 = 'PRESS' / SUM('PRESS')
MTB > NAME C4 'PRESS-R'
MTB > LET C5 = 'NOPRESS' / SUM('NOPRESS')
MTB > NAME C5 'NOPRS-R'
MTB > MPLOT 'PRESS-R' VS 'NERR' AND 'NOPRS-R' VS 'NERR'
```

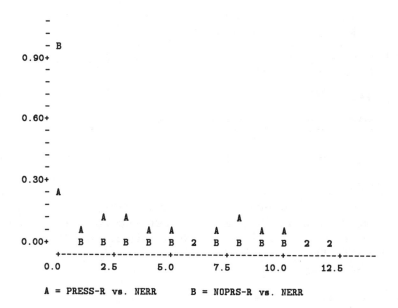

```
          -
          -
          - B
    0.90+
          -
          -
          -
          -
    0.60+
          -
          -
          -
          -
    0.30+
          - A
          -
          -         A    A                      A
          -    A              A    A       A         A    A
    0.00+    B    B    B    B    B    2    B    B    B    B    2    2
        +---------+---------+---------+---------+---------+------
        0.0      2.5      5.0      7.5     10.0     12.5

    A = PRESS-R vs. NERR      B = NOPRS-R vs. NERR
```

This plot corresponds to Figure 2.9. We could draw the bar outlines on the plot by hand up to the plotting letter to obtain the bar chart shown in the text.

2.3.2 Continuous Data

Accuracy. Minitab can represent continuous data very accurately, usually far more accurately than they were measured. This insures numerical accuracy in the results of calculations. When extracting results from Minitab for presentation, it is often appropriate to round the given values by hand to a number of significant figures comparable with the original data.

Rounding. Minitab has a function for rounding values to integers which can be applied to either a column or a constant.

ROUND E to the nearest integer, put into **C**

Numbers exactly halfway between integers (such as 4.5 or -2.5) are rounded away from zero (4.5 to 5, and -2.5 to -3). Note that this differs slightly from the rounding rule suggested in the text.

2.3.3   Frequency Distributions for Continuous Data

The HISTOGRAM command allows us to summarize continuous data into frequency data for class intervals:[3]

HISTOGRAM for **C**

with subcommands

HISTOGRAM   **START** midpoint = **K**, [end midpoint = **K**]

and

HISTOGRAM   **INTERVAL** width = **K**

If we do not specify the midpoint and interval width, Minitab will pick values which are reasonable to show the chart. For example, suppose we have entered the weights of 25 employees from Table 2.9 into C1, and named C1 'WEIGHT'. Then we obtain the following histogram:

```
MTB > HISTOGRAM 'WEIGHT'

Histogram of WEIGHT   N = 25

Midpoint   Count
     140      3  ***
     145      6  ******
     150      6  ******
     155      4  ****
     160      3  ***
     165      2  **
     170      1  *
```

Note that Minitab chose the same midpoint and interval width as in the text. By using the optional starting point and interval width, we can control how Minitab displays the histogram:

```
MTB > HISTOGRAM 'WEIGHT';
SUBC> START = 135;
SUBC> INCREMENT = 10.

Histogram of WEIGHT   N = 25

Midpoint   Count
  135.0       1   *
  145.0      11   ***********
  155.0       7   *******
  165.0       5   *****
  175.0       1   *
```

With too many small intervals the histogram will look empty and with too few big intervals the histogram will not show much detail. Usually Minitab will make a reasonable choice of intervals.

The command

DOTPLOT for C
---

will produce a horizontal plot of the values, similar to Figure 2.10.

```
MTB > DOTPLOT 'WEIGHT'

         .     .  ..  ..  .:.  :.  ..  ...   .      ...       . .          .
      ---+---------+---------+---------+---------+---------WEIGHT
      140.00    147.00    154.00    161.00    168.00
```

DOTPLOT has the same subcommands as HISTOGRAM.

Cumulative Distribution of Continuous Data. The method discussed in section 2.3.1 for plotting distribution functions works for continuous data as well:

```
MTB > NOTE CUMULATIVE RELATIVE FREQUENCY PLOT
MTB > SORT 'WEIGHT', PUT C2
MTB > NAME C2 'WEIGHT-S'
MTB > COUNT 'WEIGHT', PUT K1
   COUNT   =      25.000
MTB > GENERATE INTEGERS FROM 1 TO K1 , PUT IN C3
MTB > DIVIDE C3 BY K1, PUT IN C4
MTB > NAME C4 'RLFRQ'
MTB > PLOT 'RLFRQ' BY 'WEIGHT-S'
```

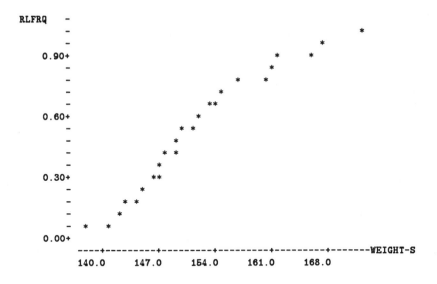

Stem-and-leaf Plots. Minitab also can print a version of the stem-and-leaf plot for a variable:

STEM-AND-LEAF display for **C**

For instance, with the employee weights data, we have

```
MTB > STEM-AND-LEAF DISPLAY OF 'WEIGHT'

Stem-and-leaf of WEIGHT   N  = 25
Leaf Unit = 1.0

    1    13 8
    5    14 1234
   12    14 5677899
   (5)   15 01234
    8    15 57
    6    16 012
    3    16 67
    1    17 2
```

This version of the stem-and-leaf is slightly different from that given in the text, but the concept is the same. Each line represents a class 5 units wide, with 14* indicating the interval 139.5 - 144.5 and 14. indicating the interval 144.5 - 149.5, etc. The parenthesized number (5) is the frequency of the middle interval (that is, the median). The numbers above are the cumulative frequencies, and the numbers below are the cumulative frequencies starting from the upper end.

APPENDIX 2B READING DATA FROM FILES

As we begin to work with somewhat larger data sets, it becomes hard to type the numbers directly into Minitab as we do our analysis without obtaining sore fingers and making typing mistakes. We would like to be able to type our numbers once and for all, check them for typing errors, correct any mistakes, and then store the data away in a form where we can access them easily for our Minitab analyses. This is done by typing the numbers into a *data file*. A *file* is just some information stored in the computer with a name on it, somewhat like a real file folder. Computer files can contain many things, and a data file is a file that has our data in it.

The particulars of how we create a file, how we name it, and how we type and correct our data in it will depend on the particular computer system we are using. An *editor* program is usually used to do these tasks. The data file we create should have the data typed into it exactly as we would type it to Minitab directly.

Once the data file has been created and the data typed into it, we can read the data into Minitab with the extended forms of the SET and READ commands:

SET data from file **'FILENAME'** into C

READ data from file **'FILENAME'** into C,...,C

The file name must be given in quotes. The exact forms that file names can take will depend on the particular type of computer system. Minitab will read the file one line at a time, just as if we were typing it in, though we shall not see the lines from the file. Identifying text information such as names can be typed into the file, and they will be ignored by Minitab when reading the file.[4] When the reading is done, Minitab will print out the first few lines of the data, and then prompt for the next command.

Most the data sets used in this *Guide* have been prepared as computer files and will be referred to in a number of chapters. Complete descriptions of the data sets as they have been set up for use in Minitab are given in the appendix. In examples and excercises where these data sets are used, the name of the disk file is given in brackets [ ].

APPENDIX 2C PRINTING MINITAB OUTPUT

Often we will want to obtain a printed record of our Minitab session. This can be done by giving the command

PAPER (put terminal output on paper)

A copy of rest of the Minitab session, including both the commands and the results, is collected and sent to the computer's printer at the end of the Minitab session. The command

NOPAPER (output to terminal only)

turns off the printing of output. The command

OUTFILE to 'FILENAME' (put output in this file)

collects the output from the session in a computer file that can be printed later using computer system commands. If the file already exists, the OUT-FILE command appends the new output to the end of the file. The command

NOOUTFILE (output to terminal only)

turns off collecting the output in the file.

APPENDIX 2D READING COMMANDS FROM FILES

It is also often useful to be able to read Minitab commands from a file, in order to replay and modify a series of analyses. The commands can be typed into a file with an editor program, as above, in the same form as they would be typed to Minitab. After preparing the command file, Minitab is started, and the command

EXECUTE commands from 'FILENAME'

reads the commands from the file and executes them exactly as they were typed. If modifications to the commands are needed, Minitab can be exited and the editor used to make the necessary changes.

Another way to place commands into a file from within Minitab is the command

STORE commands in 'FILENAME'

After giving this command the prompt changes to STOR>. Minitab does not execute the commands, but stores them in the file for future execution. The command

END of stored commands

ends the storing of commands. The EXECUTE command can then be given to execute the stored commands. If any changes are needed in the stored commands, Minitab can be exited and the editor used, as above.

COMMAND REFERENCE

Listed below are the general forms of the Minitab commands discussed in this chapter.

TABLE data in C
    COUNTS - frequency counts for each entry
    TOTPERCENTS - percentages of the total table
    FREQUENCIES in C
TALLY data in C,...,C
    COUNTS - frequency counts
    PERCENTS - relative percentages
    CUMCNTS - cumulative counts
    CUMPCTS - cumulative relative percentages
ROUND E to the nearest integer, put into **C**
HISTOGRAM for C,...,C
    START midpoint = **K** [end midpoint = **K**]
    INTERVAL width = **K**
DOTPLOT for C,...,C
    START midpoint = **K** [end midpoint = **K**]
    INTERVAL width = **K**
PLOT data in C versus data in C
MPLOT C vs C, [C vs C, C vs C, ...]
    PLOT and MPLOT have scaling subcommands:
    YINCREMENT = **K**
    YSTART at **K** [end at **K**]
    XINCREMENT = **K**
    XSTART at **K** [end at **K**]
SORT data in C, [carry along C,...,C] put in C [and C,...,C]
    The extra columns are sorted according to the order of the first column.

COUNT the number of values in **C**, put result in **K**
GENERATE integers from 1 to **K**, put in **C**
STEM-AND-LEAF display for **C**,...,**C**
SET data from file **'FILENAME'** into **C**
READ data from file **'FILENAME'** into **C**,...,**C**
PAPER (put terminal output on paper)
NOPAPER (output to terminal only)
OUTFILE 'FILENAME' (put output in this file)
NOOUTFILE (output to terminal only)
STORE commands in **'FILENAME'**
END of stored commands
EXECUTE commands from **'FILENAME'**
COPY C into **C**
    USE rows where **C = K**,...,**K**
    OMIT rows where **C = K**,...,**K**

COMPUTER EXERCISES

2.1 Using the weights and heights of Stanford football players from Table 2.16 [FOOTBALL.DAT], perform the following analyses:

(a) Tabulate the frequencies of heights, by inches.

(b) Tabulate the relative frequencies of heights, by inches.

(c) Tabulate the frequencies and relative frequencies of heights on the same table.

(d) Construct a dotplot and histogram of heights.

(e) Construct a histogram of weights, allowing Minitab to choose the interval widths. What interval width did it choose? What is the range of the first interval?

(f) Construct a histogram of weights using a larger interval width.

2.2 Demonstrate a way to use the ROUND function to round values to the nearest $\frac{1}{2}$.

2.3 For the income data from 49 families from Table 2.17 [INCOME.DAT]:

(a) Generate a histogram according to class intervals 37.5-62.5, 62.5-87.5, etc.

(b) Plot the cumulative distribution of the (ungrouped) income data. Draw lines on your output in the proper places to complete your plot.

2.4 Read in the student data from Table 2.21 [STUDENT.DAT]. The majors are coded by architecture (A) = 1, chemical engineering (CE) = 2, electrical engineering (EE) =3, industrial administration (IA) = 4, materials sciences (M) = 5, psychology (P) = 6, public administration (PA) = 7, and statistics (S) = 8. The yes/no questions are coded with no = 0, yes = 1.

(a) Construct a frequency tabulation, relative frequency tabulation, and histogram of majors.

(b) Use the COPY command[5]

```
COPY  C into C
```

with the subcommand

```
COPY
      USE rows where C = K,...,K
```

to obtain the final grades for only those students who had no previous semesters of probability and statistics. (Copy the rows of the 'Final Grade' column for which the 'Sem. Stat.' column has the value 0.) Generate a histogram of the final grade for these students. Do the same for those with at least one course in probability or statistics, using the subcommand

```
COPY
      OMIT rows where C = K,...,K
```

to select rows where 'Sem. Stat.' is not 0. Make both histograms have the same starting midpoint and interval width, so they can be compared side-by-side. Does it appear that a previous course in statistics or probability is an important factor in explaining the final grade?

2.5 Plot the time series for the number of banks in the United States from Table 2.26 with the horizontal axis for years, and the vertical axis for the number of banks.

2.6 Create a data set for the populations of California and Massachusetts from Table 2.27, with one column for the year, one column for the Massachusetts population, and one for the California population. Plot the populations for each state versus year on the same graph, with year on the horizontal axis. Draw lines between the symbols by hand, and make a line graph for each state.

NOTES

1. In the current version, the values must be integers between -10000 and 10000. In earlier versions the values must be integers between -1000 and 1000.

2. Not available in earlier versions of Minitab.

3. In the earlier versions of Minitab, the HISTOGRAM command was of the form

---
HISTOGRAM for C [first midpoint **K**, interval width **K**]

---

with no subcommands.

4. The case where there might be a problem with text in a file is when the first four letters of line match those of a Minitab command (For instance, the name SAMPson and the command SAMPle). If this happens, Minitab thinks you are done entering data and are giving a new command, and will probably be confused. In such cases putting the identifying information after the numbers on each line will avoid the problem.

5. In earlier versions of Minitab, the command for selecting data is

---
CHOOSE rows with the value **K** [through **K**] in C, corresp. rows of C,...,C put in C and C,...,C

---

3 MEASURES OF LOCATION

INTRODUCTION

In this chapter we shall see how to use Minitab to compute measures of location (central tendency) and percentiles for a set of data.

3.1 THE MODE

3.1.1 Definition and Interpretation

The mode is the value in the data which occurs most frequently. There is no predefined command in Minitab to calculate the mode, but for discrete data calculating a histogram makes finding the mode quite easy. For instance, using the data from Table 2.6 on numbers of children in 20 households, we have

```
MTB > SET C1
DATA> 2 0 1 0 0 1 1 1 4 1 3 2 2 1 1 2 0 3 1 4
DATA> END OF DATA
MTB > NAME C1 'NCHILD'
MTB > HISTOGRAM FOR 'NCHILD'

Histogram of NCHILD    N = 20

Midpoint   Count
       0       4   ****
       1       8   ********
       2       4   ****
       3       2   **
       4       2   **
```

The modal number of children in this survey is 1.

3.1.2 Mode of Grouped Data

The HISTOGRAM command is also effective for grouping continuous data to find a mode. Minitab does a pretty good job of picking interval widths; alternatively you can choose your own class definitions. For example, using the income data for 49 families from Table 2.17, we have

```
MTB > NOTE INCOME DATA FOR 49 FAMILIES - TABLE 2.17
MTB > READ DATA FROM 'INCOME.DAT' INTO C1 C2
     49 ROWS READ
  ROW      C1      C2

    1    31031     85
    2    31059    135
    3    31069    200
    4    31159     50
    .   .   .

MTB > NAME C1 'RESPNUM' C2 'INCOME'
MTB > HISTOGRAM 'INCOME';
SUBC> STARTING MIDPOINT = 30;
SUBC> INCREMENT =  20.

Histogram of INCOME   N = 49

Midpoint    Count
   30.0       0
   50.0       2  **
   70.0      16  ****************
   90.0      11  ***********
  110.0      10  **********
  130.0       7  *******
  150.0       2  **
  170.0       0
  190.0       0
  210.0       1  *
```

With respect to these class definitions, the modal interval is from $6000–$8000; we can take the mode as $7000.

3.2 THE MEDIAN AND OTHER PERCENTILES

3.2.1 The Median

The MEDIAN command calculates the median value in a set of data, defined as the middle of the ordered sample if there is an odd number of cases and the average of the two middle values if there is an even number of cases.

MEDIAN of the values in C

For instance, with the weights of 58 swine from Table 2.18, we have

```
MTB > SET DATA FROM 'SWINE.DAT' INTO C1
C1
    73    139    143    195     .    .    .
```

```
MTB > NAME C1 'WEIGHT'
MTB > MEDIAN OF 'WEIGHT'
   MEDIAN =       139.00
```

The median weight of the swine is 139 pounds.

3.2.2   Quartiles

There is no special command in Minitab to compute quartiles. However, the command

DESCRIBE data in C

prints a number of useful descriptive statistics on a data column, including the first and third quartiles (Q1 and Q3). For instance:

```
MTB > DESCRIBE 'WEIGHT'
```

	N	MEAN	MEDIAN	TRMEAN	STDEV	SEMEAN
WEIGHT	58	149.64	139.00	146.21	58.87	7.73

	MIN	MAX	Q1	Q3
WEIGHT	64.00	298.00	105.25	184.50

The first quartile of swine weights is 105.25 pounds, and the third quartile is 184.50 pounds.

3.2.3   Deciles, Percentiles, and Other Quantiles

There is no command in Minitab to calculate other quantiles directly. However, if we plot the cumulative distribution plot of the data, as described in Chapter 2, we can find the quantiles on the plot. For instance, if we want the first and ninth deciles of the swine weights, we have

```
MTB > SORT 'WEIGHT', PUT IN C2
MTB > NAME C2 'WGT-S'
MTB > COUNT 'WEIGHT', PUT IN K1
   COUNT    =      58.000
MTB > GENERATE 1 THROUGH K1, PUT IN C3
MTB > LET C4 = C3/K1
MTB > NAME C4 'DIST'
MTB > PLOT 'DIST' VS 'WGT-S'
```

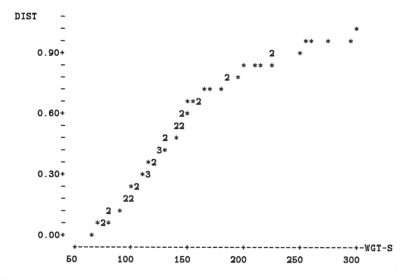

Drawing horizontal lines at the 0.10 and 0.90 level gives values of approximately 75 for the first decile and 250 for the ninth decile of swine weights. To obtain more exact values, we can print out the cumulative distribution function.

```
MTB > PRINT 'DIST','WGT-S'
 ROW     DIST  WGT-S

   1   0.01724    64
   2   0.03448    72
   3   0.05172    73
   4   0.06897    73
   5   0.08621    78
   6   0.10345    78
   7   0.12069    82
   8   0.13793    89
   9   0.15517    97
  10   0.17241    97
  11   0.18966   100
  12   0.20690   102
  13   0.22414   102
  14   0.24138   103
  15   0.25862   106
  16   0.27586   109
  17   0.29310   113
  18   0.31034   115
  19   0.32759   115
  20   0.34483   117
  21   0.36207   118
  22   0.37931   119
  23   0.39655   126
```

24	0.41379	126
25	0.43103	126
26	0.44828	130
27	0.46552	131
28	0.48276	132
29	0.50000	139
30	0.51724	139
31	0.53448	141
32	0.55172	143
33	0.56897	144
34	0.58621	144
35	0.60345	145
36	0.62069	151
37	0.63793	152
38	0.65517	156
39	0.67241	160
40	0.68966	160
41	0.70690	166
42	0.72414	172
43	0.74138	179
44	0.75862	184
45	0.77586	186
46	0.79310	195
47	0.81034	198
48	0.82759	211
49	0.84483	214
50	0.86207	224
51	0.87931	224
52	0.89655	227
53	0.91379	251
54	0.93103	256
55	0.94828	258
56	0.96552	273
57	0.98276	296
58	1.00000	298

The first row with DIST > 0.10 is row 6 so the first decile is 78. The first row with DIST > 0.90 is row 53, so the ninth decile is 251.

In Chapter 8 we shall discuss a way to obtain quantiles using a series of Minitab commands, without having to print out the cumulative distribution function.

3.3   THE MEAN

3.3.1   Definition and Interpretation

The mean or average value of a set of data can be calculated with the MEAN command:

MEAN of the values in C

For the swine weights:

```
MTB > MEAN 'WEIGHT'
   MEAN    =      149.64
```

3.3.2   Use of Notation

As we have seen, the mean can be written in summation notation as

$$\bar{x} = \frac{1}{n} \sum_{i=1}^{n} x_i.$$

To use this formula, we calculate the sum of the values in a column with the command

SUM of the values in **C**, put result in **K**

Then dividing by the sample size n, which we can obtain from the

COUNT the values in **C**, put result in **K**

command, we have a different way to calculate the mean. For instance,

```
MTB > SUM 'WEIGHT', PUT IN K1
   SUM     =       8679.0
MTB > COUNT 'WEIGHT', PUT IN K2
   COUNT   =       58.000
MTB > LET K3 = K1/K2
MTB > PRINT K3
K3       149.638
```

And since both the SUM and COUNT functions are available as part of the LET command, we can write:

```
MTB > LET K4 = SUM('WEIGHT')/COUNT('WEIGHT')
MTB > PRINT K4
K4       149.638
```

3.3.3   Calculating the Mean of Discrete Data

Our exercise in the preceeding section becomes more useful when we need to calculate the mean of discrete data based on frequency tabulations. If we have the values v_j in one column, and the frequencies f_j in another, we compute according to the formula

$$\bar{x} = \frac{1}{n} \sum_{j=1}^{m} f_j v_j,$$

where

$$n = \sum_{j=1}^{m} f_j.$$

This is computed here with data on the number of children in 10 families from Table 3.7.

```
MTB > READ C1 C2 C3
DATA> 1 2 1
DATA> 2 3 5
DATA> 3 4 3
DATA> 4 6 1
DATA> END OF DATA
       4 ROWS READ
MTB > NAME C1 'INDEX' C2 'CHILDREN' C3 'FREQ'
MTB > LET K1 = SUM('FREQ'*'CHILDREN')/SUM('FREQ')
MTB > PRINT K1
K1        3.50000
```

3.3.4   Calculating the Mean of Grouped Data

The same technique can be applied to calculating the mean of grouped data. Here we have the grouped data on the incomes of 64 families from Table 3.10:

```
MTB > READ C1 C2 C3
DATA> 1  2000   4
DATA> 2  6000   6
DATA> 3 10000  15
DATA> 4 14000  28
DATA> 5 18000   7
DATA> 6 22000   2
DATA> 7 26000   2
DATA> END OF DATA
       7 ROWS READ
MTB > NAME C1 'INDEX' C2 'INCOME' C3 'FREQ'
MTB > LET K1 = SUM('INCOME'*'FREQ')/SUM('FREQ')
MTB > PRINT K1
K1       12625.0
```

3.3.5   Means of Indicator Variables

As discussed in the text, the mean of a variable that only takes on the values 0 and 1 is equal to the proportion of the cases with 1's. We can use this to find the proportion of data satisfying a criterion. For instance if we have the occupational data from Table 2.1 coded as we did in Chapter 2, and we wish to find the proportion of people with a professional occupation, we create a new column with 1's for the professionals and 0's elsewhere using the command[1]

CODE (K,...,K) to K for C, put result in C

Since professionals were coded with a 1, it suffices to change all other values (2 to 4) to 0. In the parenthesized list of values to change, a range of values can be indicated with a colon, as shown below:

```
MTB > CODE VALUES (2:4) TO 0 IN 'OCCUP', PUT IN C2
MTB > NAME C2 'PROF'
MTB > PRINT 'OCCUP' 'PROF'
 ROW  OCCUP   PROF

   1      3      0
   2      1      1
   3      1      1
   4      1      1
   5      3      0
   6      1      1
   7      1      1
   8      2      0
   9      4      0
  10      2      0
  11      2      0
  12      2      0
  13      2      0
  14      2      0
  15      1      1
  16      3      0
  17      2      0
  18      3      0
  19      2      0
  20      4      0
```

The 'PROF' is 1 if and only if 'OCCUP' is 1, and is 0 otherwise. The mean of 'PROF' is the desired proportion.

```
MTB > MEAN 'PROF'
  MEAN    =      0.30000
```

So 0.3 of the subjects (30%) are professionals.

3.4 WEIGHTED AVERAGES AND STANDARDIZED AVERAGES

Weighted and standardized averages are computed exactly as with frequency data above, by multiplying a weighting factor by the value for each group, obtaining the sum, and dividing by the total weight.

COMMAND REFERENCE

Listed below are the general forms of the Minitab commands discussed in this chapter.

MEDIAN of the values in C [put result in **K**]

DESCRIBE data in C,...,C

MEAN of the values in C [put result in **K**]

SUM of the values in **C** [put result in **K**]
COUNT the values in **C**, put result in **K**
CODE (**K**,...,**K**) to **K** for **C**, put result in **C**

COMPUTER EXERCISES

3.1 For the heights and weights of Stanford football players from Table 2.16 [FOOTBALL.DAT]:

(a) Generate histograms of the heights and weights.

(b) Compute the mean, median, and mode of the heights and weights.

(c) Generate the cumulative distribution of the weights and compute the first and third quartiles.

(d) Generate the cumulative distribution of the heights and estimate the first decile and 9th decile.

3.2 Create a data set with the reaction times in Table 3.27 with a separate column for each subject. Compute mean and median reaction times for each subject.

3.3 Use the farm data set in Table 3.20 [FARM.DAT]. Compute the average number of toys *per child*.

3.4 Use income data for 49 families from Table 2.17 [INCOME.DAT].

(a) Compute the mean income.

(b) Obtain a histogram of the incomes with first midpoint 30 and interval width 20. Create a new data set with grouped income frequencies using the information from the histogram.

(c) Compute the mean income using the grouped data. How close is it to the value obtained using the ungrouped data?

3.5 Compute the mean and median light bulb lifetime for the data in [BULB.DAT]. Obtain a histogram, and comment on the shape of the distribution.

NOTES

1. In earlier versions of Minitab the command to recode data is

RECODE values from **K** [to **K**] in **C** to **K**, put result in **C**

4 MEASURES OF VARIABILITY

INTRODUCTION

In this chapter we shall see how to use Minitab to compute measures of variability.

4.1 RANGES

4.1.1 The Range

The sample range is defined as the difference between the largest observation (maximum) and smallest observation (minimum). We can compute the range using Minitab in exactly this way, using the MINIMUM and MAXIMUM commands:

MINIMUM of the values in C

MAXIMUM of the values in C

For example, we use the income data for 49 families from Table 2.17, which have been stored in a file called "INCOME.DAT" with each line containing the respondent number and income (in $100's) for a family. Then to compute the range, we do as follows:

```
MTB > READ DATA FROM 'INCOME.DAT' INTO C1 C2
      49 ROWS READ
  ROW     C1       C2

    1    31031       85
    2    31059      135
    3    31069      200
    4    31159       50
     .     .    .

MTB > NAME C1 'RESPNUM' C2 'INCOME'
MTB > MAXIMUM 'INCOME'
      MAXIMUM =      200.00
MTB > MINIMUM 'INCOME'
      MINIMUM =      50.000
```

```
MTB > LET K1 = MAXIMUM('INCOME') - MINIMUM('INCOME')
MTB > PRINT K1
K1        150.000
```

Therefore the range of incomes is $15,000.

4.1.2   The Interquartile Range

The interquartile range of a set of numbers is defined as the difference of the upper quartile and the lower quartile. As discussed in Chapter 3, we can use the DESCRIBE command to compute the upper and lower quartiles and then take the difference to find the interquartile range.

Using the income data as above, we proceed as follows:

```
MTB > DESCRIBE 'INCOME'

                N      MEAN    MEDIAN   TRMEAN    STDEV    SEMEAN
INCOME         49     93.65     92.00    91.87    28.96      4.14

              MIN       MAX        Q1       Q3
INCOME      50.00    200.00     70.00   107.50
```

The first quartile is at 70 and the third quartile is at 107.5. Therefore the interquartile range of incomes is $3750.

4.2   THE MEAN DEVIATION

The (population) mean deviation, defined as the mean absolute deviation of the data from their mean, can be computed by the definition. For instance, with the income data we have

```
MTB > MEAN 'INCOME' , PUT IN K1
   MEAN    =      93.653
MTB > SUBTRACT K1 FROM 'INCOME', PUT RESULT IN C3
MTB > ABSOLUTE VALUE OF C3, PUT IN C4
MTB > MEAN OF C4
   MEAN    =      22.122
```

This gives a mean deviation of incomes of $2,212, appropriate if the data are considered as constituting a population. A slightly more concise method is to use the LET command with the MEAN and ABSO(lute value) functions, as follows:

```
MTB > LET K1 = MEAN(ABSO('INCOME' - MEAN('INCOME')))
MTB > PRINT K1
K1        22.1216
```

Sample Mean Deviation. Note that both of the above methods divide by n because the MEAN function divides by n. To obtain the *sample* mean deviation

$$\frac{\sum_{i=1}^{n} |x_i - \overline{x}|}{n-1},$$

we use the LET command as above, but with the SUM and COUNT functions:

```
MTB > LET K2 = SUM(ABSO('INCOME' - MEAN('INCOME')))/(COUNT('INCOME')-1)
MTB > PRINT K2
K2      22.5825
```

Therefore the sample mean deviation of income is $2,258.

4.3   THE STANDARD DEVIATION

Because the sample standard deviation is a very commonly calculated measure of variability, Minitab has a built-in command to calculate it, with two equivalent forms:

STANDARD deviation of C

or

STDEV of C

These calculate the sample standard deviation

$$\frac{\sum_{i=1}^{n} (x_i - \overline{x})^2}{n-1}.$$

For our income example, to obtain the sample standard deviation of the income data we can do either of:

```
MTB > STANDARD DEVIATION OF 'INCOME'
   ST.DEV. =      28.959
MTB > STDEV 'INCOME'
   ST.DEV. =      28.959
```

Either way, we find that the sample standard deviation of incomes is $2,896.

Population Standard Deviation. If we were working with measurements from a complete population, we simply multiply the above result by $\sqrt{(N-1)/N}$ to obtain the correct population formula, as follows:

```
MTB > LET K1 = COUNT('INCOME')
MTB > LET K2 = SQRT((K1-1)/K1)*STDEV('INCOME')
MTB > PRINT K2
K2      28.6624
```

For the 49 families considered as a population, the population standard deviation is $2,866.

Sample Variance. Because the sample variance is just the square of the sample standard deviation, we can easily compute it as follows:

```
MTB > LET K1 = STDEV('INCOME')**2
MTB > PRINT K1
K1        838.648
```

The sample variance of incomes is 838.648 in $100's squared, or 838,648 dollars squared.

Population Variance. As above, the results of the population formula for variance can be easily obtained. Here we multiply the square of the standard deviation by $(N-1)/N$:

```
MTB > LET K1 = COUNT('INCOME')
MTB > LET K2 = (K1-1)/K1*STDEV('INCOME')**2
MTB > PRINT K2
K2        821.533
```

4.3.1   Standard Deviation for Discrete Data

We can apply the formulas given in the text to calculate the standard deviation of discrete or grouped data. For instance, for the data on the number of children in 10 families from Table 4.5 we have

```
MTB > READ C1 C2 C3
DATA> 1 2 1
DATA> 2 3 5
DATA> 3 4 3
DATA> 4 6 1
DATA> END OF DATA
      4 ROWS READ
MTB > NAME C1 'INDEX' C2 'CHILDREN' C3 'FREQ'
MTB > LET K1 = SUM('FREQ')
MTB > LET K2 = SUM('FREQ'*'CHILDREN')
MTB > LET K3 = SUM('FREQ'*'CHILDREN'**2)
MTB > LET K4 = SQRT((K3 - K2**2/K1)/(K1-1))
MTB > PRINT K4
K4        1.08012
```

The standard deviation of the number of children in these families is 1.08.

COMMAND REFERENCE

Listed below are the general forms of the Minitab commands discussed in this chapter.

MINIMUM of the values in **C** [put in **K**]
MAXIMUM of the values in **C** [put in **K**]

DESCRIBE the data in columns C,...,C
STANDARD deviation of C [put in **K**]
STDEV of C [put in **K**]

COMPUTER EXERCISES

4.1 For the weights of the members of the 1970 Stanford football team from Table 2.16 [FOOTBALL.DAT] calculate:

(a) The sample standard deviation.

(b) The sample mean deviation.

(c) The interquartile range.

4.2 For the head length measurements and the head breadth measurements from Table 4.10 [HEAD.DAT] calculate:

(a) The range and interquartile range.

(b) The sample mean deviation and sample standard deviation.

(c) The *coefficient of variation*, or *relative standard deviation*, defined as the ratio of the standard deviation to the mean.

Which of head length or head breadth varies more, in actual terms? Which varies more in relative terms?

4.3 Calculate the standard deviation of the number of children in 100 families using the data in Table 3.10.

4.4 Using the grouped data for the income of 49 families as constructed in Exercise 3.4, compute the standard deviation of the grouped incomes. Compare this with the standard deviation of the ungrouped incomes.

4.5 Compute the range, interquartile range, mean absolute deviation, and standard deviation of the light bulb lifetimes in [BULB.DAT].

5 ORGANIZATION OF MULTIVARIATE DATA

INTRODUCTION

In this chapter we shall see how to use Minitab to examine the relationship between data on two or more variables for each individual. The TABLES command can be used in many ways to examine and summarize multivariate categorical data. Graphical methods are presented for examining several numerical variables.

5.1 TWO-BY-TWO FREQUENCY TABLES

5.1.1 Organization of Data

Bivariate Categorical Data. As discussed in Chapter 2, categorical data will need to be assigned numerical codes for use in Minitab. For instance, for the double dichotomy data in Table 5.1 we can assign a coding for sex of female (F) = 1 and male (M) = 2 and a coding for university division of graduate (G) = 1 and undergradate (U) = 2. Then we create the data set as follows:

```
MTB > NOTE SEX: F=1 M=2
MTB > NOTE UNIVERSITY DIVISION: G=1 U=2
MTB > READ C1 C2
DATA> 2 1            DATA> 1 1
DATA> 2 2            DATA> 2 1
DATA> 1 1            DATA> 2 2
DATA> 2 1            DATA> 2 2
DATA> 2 2            DATA> 2 2
DATA> 2 2            DATA> 2 2
DATA> 2 2            DATA> 1 1
DATA> 2 2            DATA> 1 2
DATA> 2 2            DATA> 1 2
DATA> 1 2            DATA> 1 1
DATA> 1 1            DATA> 1 2
DATA> 1 1            DATA> 1 1
DATA> 1 1            DATA> END OF DATA
     25 ROWS READ
MTB > NAME C1 'SEX' C2 'DIVISION'
```

Of course, it is usually better to type the data into a file, and read it into Minitab from the file, in case we want to perform further analyses with it.

   2×2 *Frequency Tables.* The TABLE command can be used to obtain cross-tabulations of two categorical variables.

TABLE C by C

The first mentioned column becomes the rows of the table, and the second becomes the columns. For the example above, we have

```
MTB > TABLE 'SEX' 'DIVISION'

 ROWS: SEX      COLUMNS: DIVISION

            1        2       ALL

   1        8        4        12
   2        3       10        13
 ALL       11       14        25


 CELL CONTENTS --
                  COUNT
```

This gives the frequencies obtained by hand in Table 5.2.

5.1.2   Calculation of Percentages

   Marginal Totals. As can be seen in the above example, the TABLE command automatically calculates the marginal totals for the table.

   Percentages Based on Row Totals. To find percentages based on the row totals, we use the TABLE subcommand

---

TABLE

   ROWPERCENT - percentages based on row totals

---

as shown below.

```
MTB > TABLE 'SEX' 'DIVISION';
SUBC> ROWPERCENTS.

 ROWS: SEX      COLUMNS: DIVISION

            1        2       ALL

   1      66.67    33.33    100.00
   2      23.08    76.92    100.00
 ALL      44.00    56.00    100.00


 CELL CONTENTS --
                  % OF ROW
```

Note that below the table the contents of each cell are identified.

Percentages Based on Column Totals. Similarly, column percentages are obtained with the TABLE subcommand

```
TABLE
    COLPERCENT - percentages based on column totals
```

In our example, we have

```
MTB > TABLE 'SEX' 'DIVISION';
SUBC> COLPERCENTS.

  ROWS: SEX     COLUMNS: DIVISION

              1        2      ALL

    1      72.73    28.57    48.00
    2      27.27    71.43    52.00
  ALL     100.00   100.00   100.00

  CELL CONTENTS --
              % OF COL
```

If we want to have both the frequency counts and percentages in each cell, we can use the subcommand

```
TABLE
    COUNTS - frequency count
```

in conjunction with the other subcommands. Don't forget when using subcommands to put a semicolon (;) at the end of the main command and each subcommand except the last, which ends with a period. For instance, if we want both frequency counts and column percentages in our example, we do as follows:

```
MTB > TABLE 'SEX' 'DIVISION';
SUBC> COUNT;
SUBC> COLPERCENTS.

  ROWS: SEX     COLUMNS: DIVISION

              1        2      ALL

    1          8        4       12
           72.73    28.57    48.00

    2          3       10       13
           27.27    71.43    52.00

  ALL         11       14       25
          100.00   100.00   100.00
```

```
CELL CONTENTS --
               COUNT
               % OF COL
```

Percentages Based on the Grand Total.   The TABLE subcommand

TABLE

TOTPERCENTS - percentages of two-way table
---

is also available for total percentages:

```
MTB > TABLE 'SEX' 'DIVISION';
SUBC> TOTPERCENTS.

 ROWS: SEX      COLUMNS: DIVISION

              1        2       ALL

    1      32.00    16.00    48.00
    2      12.00    40.00    52.00
  ALL      44.00    56.00   100.00

    CELL CONTENTS --
                  % OF TBL
```

5.1.3   Interpretation of Frequencies

In many cases data will come already summarized according to its frequency for each combination of variables. This actually makes the data easier to work with since we do not have to type in each individual case. Instead we set up a variable to contain the frequency information and use the TABLE subcommand

---
TABLE

FREQUENCIES in C
---

to let Minitab know that we have frequency information and where it is. Below is an example of how we proceed if we are given just the frequency information in Table 5.12 on the results of the X Company's survey, and were asked to find the percentages using Brand X blades:

```
MTB > NOTE RAZOR (C1) USE BRAND X RAZOR=1 DO NOT USE BRAND X RAZOR=2
MTB > NOTE BLADES (C2) USE BRAND X BLADES=1 DO NOT USE BRAND X BLADES=2
MTB > NOTE FREQ (C3) FREQUENCY OF RESPONSE
MTB > READ C1 C2 C3
DATA> 1 1 186
DATA> 1 2  93
DATA> 2 1  59
DATA> 2 2 262
```

```
DATA> END OF DATA
      4 ROWS READ
MTB > NAME C1 'RAZOR' C2 'BLADES' C3 'FREQ'
MTB > TABLES 'RAZOR' 'BLADES';
SUBC> FREQUENCIES 'FREQ';
SUBC> COUNTS;
SUBC> ROWPERCENTS.
```

```
 ROWS: RAZOR      COLUMNS: BLADES

              1        2      ALL

  1         186       93      279
          66.67    33.33   100.00

  2          59      262      321
          18.38    81.62   100.00

ALL         245      355      600
          40.83    59.17   100.00

  CELL CONTENTS --
                     COUNT
                     % OF ROW
```

Measures of Association. To calculate the ϕ measure of association we use the TABLE subcommand

---

TABLE

CHISQUARE - calculate chi-squared statistic

---

which calculates a statistic known as the chi-squared (X^2) statistic. (We shall learn more about the X^2 statistic in Chapter 12.) For 2×2 tables it is related to the ϕ statistic by

$$X^2 = n\phi^2.$$

So we can calculate the absolute value of ϕ as $\sqrt{X^2/n}$. Then ϕ is positive if $ad > bc$ and is negative if $ad < bc$. Here is how it works in our razor blade example:

```
MTB > TABLES 'RAZOR' 'BLADES';
SUBC> FREQUENCIES 'FREQ';
SUBC> CHISQUARE.
```

```
 ROWS: RAZOR      COLUMNS: BLADES

              1        2      ALL

  1         186       93      279
```

```
       2      59     262     321
     ALL     245     355     600

  CHI-SQUARE =    144.052   WITH D.F. =    1

     CELL CONTENTS --
                  COUNT

  MTB > LET K2 = SQRT(142.060/SUM('FREQ'))
  MTB > PRINT K2
  K2       0.486587
```

Here the sign of ϕ is not important because the levels of the variables are only nominal, and so we have $\phi = 0.487$ as our index of association, which indicates fairly strong association between using brand X razors and brand X blades.

In cases where the levels of the variables are ordered, we can usually determine the sign of ϕ by inspection. For instance, with the social status and income data from Table 5.19 we would have

```
  MTB > NOTE LOW=1, HIGH=2
  MTB > READ C1 C2 C3
  DATA> 1 1 100
  DATA> 1 2  25
  DATA> 2 1  50
  DATA> 2 2  75
  DATA> END OF DATA
        4 ROWS READ
  MTB > NAME C1 'INCOME' C2 'SOCSTAT' C3 'FREQ'
  MTB > TABLE 'INCOME' 'SOCSTAT';
  SUBC> FREQUENCIES 'FREQ';
  SUBC> CHISQUARE.

   ROWS: INCOME     COLUMNS: SOCSTAT

              1       2     ALL

      1     100      25     125
      2      50      75     125
    ALL     150     100     250

  CHI-SQUARE =    41.667   WITH D.F. =    1

     CELL CONTENTS --
                  COUNT

  MTB > LET K1 = SQRT(40.017/SUM('FREQ'))
  MTB > PRINT K1
  K1       0.400085
```

Here we can see positive association and ϕ has a positive sign. With the grade-point average data from Table 5.20 we have

```
MTB > NOTE LOW=1, HIGH=2
MTB > READ C1 C2 C3
DATA> 1 1 200
DATA> 1 2  40
DATA> 2 1 250
DATA> 2 2  10
DATA> END OF DATA
      4 ROWS READ
MTB > NAME C1 'GPA' C2 'SEMSTAT' C3 'FREQ'
MTB > TABLE 'GPA' 'SEMSTAT';
SUBC> FREQUENCIES 'FREQ';
SUBC> CHISQUARE.

   ROWS: GPA      COLUMNS: SEMSTAT

              1        2      ALL

      1     200       40      240
      2     250       10      260
    ALL     450       50      500

CHI-SQUARE =     22.792   WITH D.F. =    1

   CELL CONTENTS --
                    COUNT

MTB > LET K1 = SQRT(21.390/SUM('FREQ'))
MTB > PRINT K1
K1       0.206833
```

Here the association is negative and $\phi = -0.207$.

5.2 LARGER TWO-WAY FREQUENCY TABLES

5.2.1 Organization of Data for Two Categorical Variables

All of the commands discussed above work for larger two-way tables. The size of the table is determined by the number of different integer values taken on by the codes in each variable.[1] For instance, with the eye and hair color data from Table 5.21 we have

```
MTB > NOTE EYE (C1) BLUE=1 GREY OR GREEN=2 BROWN=3
MTB > NOTE HAIR (C2) FAIR=1 BROWN=2 BLACK=3 RED=4
MTB > NOTE FREQ (C3) FREQUENCIES
MTB > READ C1 C2 C3
DATA> 1 1 1768
DATA> 1 2  807
DATA> 1 3  189
DATA> 1 4   47
DATA> 2 1  946
```

```
DATA> 2 2 1387
DATA> 2 3  746
DATA> 2 4   53
DATA> 3 1  115
DATA> 3 2  438
DATA> 3 3  288
DATA> 3 4   16
DATA> END OF DATA
     12 ROWS READ
MTB > NAME C1 'EYE' C2 'HAIR' C3 'FREQ'
MTB > TABLES 'EYE' BY 'HAIR';
SUBC> FREQUENCIES IN 'FREQ';
SUBC> COUNTS;
SUBC> ROWPERCENTS;
SUBC> COLPERCENTS;
SUBC> TOTPERCENTS.


 ROWS: EYE    COLUMNS: HAIR

              1        2        3        4      ALL

    1      1768      807      189       47     2811
           62.90    28.71     6.72     1.67   100.00
           62.50    30.66    15.45    40.52    41.34
           26.00    11.87     2.78     0.69    41.34


    2       946     1387      746       53     3132
           30.20    44.28    23.82     1.69   100.00
           33.44    52.70    61.00    45.69    46.06
           13.91    20.40    10.97     0.78    46.06


    3       115      438      288       16      857
           13.42    51.11    33.61     1.87   100.00
            4.07    16.64    23.55    13.79    12.60
            1.69     6.44     4.24     0.24    12.60


  ALL      2829     2632     1223      116     6800
           41.60    38.71    17.99     1.71   100.00
          100.00   100.00   100.00   100.00   100.00
           41.60    38.71    17.99     1.71   100.00


   CELL CONTENTS --
                    COUNT
                  % OF ROW
                  % OF COL
                  % OF TBL
```

5.3   THREE CATEGORICAL VARIABLES

5.3.1   Organization of Data for Three Yes-No Variables

The TABLE command can also be used for examining observations on more than two categorical variables.

TABLE C by C by C ...

When such a table is requested a two way table of the first two columns is produced for each level of the other variables. For instance with the data from Table 5.32 on use of razors, blades, and commercial viewing we have

```
MTB > NOTE RAZOR (C1) USE BRAND X RAZOR=1 NOT USE BRAND X RAZOR=2
MTB > NOTE BLADES (C2) USE BRAND X BLADES=1 NOT USE BRAND X BLADES=2
MTB > NOTE COMMRCL (C3) HAVE SEEN COMMERCIAL=1 NOT SEEN COMMERCIAL=2
MTB > NOTE FREQ (C4) FREQUENCIES
MTB > READ C1-C4
DATA> 1 1 1  86
DATA> 1 1 2 100
DATA> 1 2 1  38
DATA> 1 2 2  55
DATA> 2 1 1   9
DATA> 2 1 2  50
DATA> 2 2 1  17
DATA> 2 2 2 245
DATA> END OF DATA
      8 ROWS READ
MTB > NAME C1 'RAZOR' C2 'BLADES' C3 'COMMRCL' C4 'FREQ'
MTB > TABLES 'RAZOR' BY 'BLADES' BY 'COMMRCL';
SUBC> FREQUENCIES 'FREQ'.

CONTROL: COMMRCL = 1
ROWS: RAZOR      COLUMNS: BLADES

            1         2      ALL

   1        86        38      124
   2         9        17       26
 ALL        95        55      150

CONTROL: COMMRCL = 2
ROWS: RAZOR      COLUMNS: BLADES

            1         2      ALL

   1       100        55      155
   2        50       245      295
 ALL       150       300      450

   CELL CONTENTS --
                 COUNT
```

The third variable is called the *control* variable. We can think of it as representing the third dimension of the table.

Table Layout. If we have several control variables, the number of tables printed can become quite unwieldy. Even in our example we would like to produce a table like Table 5.32. We can do this with the TABLE subcommand

---

TABLE

LAYOUT with first **K** variables for rows, next **K** for columns

---

The default layout is 1,1 (one row variable and one column variable, and any other variables as control variables). If in our example we reorder the list of columns and specifiy 1 row and 2 column variables, we have

```
MTB > TABLES 'RAZOR' BY 'COMMRCL' BY 'BLADES';
SUBC> FREQUENCIES 'FREQ';
SUBC> LAYOUT 1,2.
```

ROWS: RAZOR   COLUMNS: COMMRCL / BLADES

	1		2		ALL
	1	2	1	2	ALL
1	86	38	100	55	279
2	9	17	50	245	321
ALL	95	55	150	300	600

```
       CELL CONTENTS --
               COUNT
```

This is much more readable. Note however in such layouts row and column statistics refer to the entire row or column of the table as printed.

Marginal Tables. If we have three categorical variables, but only tabulate according to two of them, we obtain the result of summing the three-way tables over the third variable. These are the *two-way marginal tables* of the three way table. For instance, in our example we have

```
MTB > TABLES 'RAZOR' BY 'BLADES';
SUBC> FREQUENCIES 'FREQ'.
```

ROWS: RAZOR   COLUMNS: BLADES

	1	2	ALL
1	186	93	279
2	59	262	321
ALL	245	355	600

```
CELL CONTENTS --
            COUNT
```

5.3.2   Larger Three-way Frequency Tables

Three-way and larger frequency tables with more than two categories per variable are handled exactly as above.

5.4   EFFECTS OF A THIRD VARIABLE

The exploration of association between variables and the effects of a third variable can be explored using the TABLE command as described in the preceeding sections.

5.5   SEVERAL NUMERICAL VARIABLES

5.5.1   Scatter Plots

We can examine the relationship between several numerical variables with scatter plots using the command

PLOT data in C versus data in C

The first mentioned variable is plotted on the y (vertical) axis, and the second mentioned variable is plotted on the x (horizontal) axis. For instance, with the intellegence data from Table 2.23 [IQ.DAT] we can plot the relationship between nonlanguage IQ and language IQ as follows:

```
MTB > READ 'IQ.DAT' C1-C4
      23 ROWS READ
  ROW    C1      C2      C3      C4

    1     1      86      94     1.1
    2     2     104     103     1.5
    3     3      86      92     1.5
    4     4     105     100     2.0
     .     .     .

MTB > NAME C1 'PUPIL' C2 'LNGIQ' C3 'NONLNGIQ'
MTB > NAME C4 'INITREAD' C5 'FINLREAD'
MTB > NOTE SCATTER PLOT
MTB > PLOT 'NONLNGIQ' AS Y VERSUS 'LNGIQ' AS X
```

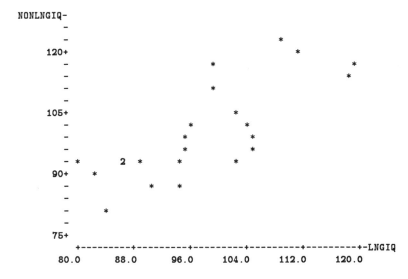

Sometimes we would like to be able to go back to the data and look more closely at an individual point. A plot with different symbols for each point provides a convenient way to find the point in the data set. The command

---
LPLOT C vs C, labeled by C
---

uses the values in the third variable to assign a different symbol to each point in the plot. The value 1 in the label column is plotted as 'A', 2 as 'B', etc. If the value in the label column is greater than 26 the letters are repeated, that is 27 = 'A', 28 = 'B', etc. This version of the plot above is shown on the next page.

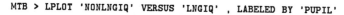

```
MTB > LPLOT 'NONLNGIQ' VERSUS 'LNGIQ' , LABELED BY 'PUPIL'
```

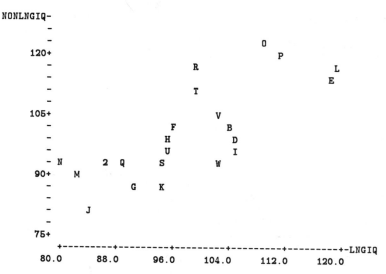

```
NONLNGIQ-
       -
       -                                          O
  120+                                         P
       -                    R
       -                                           L
       -                    T                      E
       -
  105+                          V
       -               F       B
       -               H       D
       -               U       I
     - N       2  Q    S       W
   90+  M
       -           G  K
       -
       -     J
       -
   75+
       +---------+---------+---------+---------+---------+-LNGIQ
       80.0      88.0      96.0     104.0     112.0     120.0
```

5.5.2 Descriptive Statistics for Bivariate Numerical Data; Correlation Coefficient

The correlation coefficient is calculated by the command

CORRELATION of C and C

For instance, to obtain the correlation of the language and non-language IQ's in the above example, we do

```
MTB > CORRELATION OF 'NONLNGIQ' WITH 'LNGIQ'

Correlation of NONLNGIQ and LNGIQ = 0.769
```

5.5.3 Time Plots

Another type of bivariate relationship is that of a variable with time. For instance, [AIRLINE.DAT] contains monthly commercial air passenger totals (in thousands of passengers) for the period January 1949 to December 1960. We can use the command

---
TSPLOT [with period **K**] time series data in C
---

to examine the series. Because the data are monthly, we specify a period of 12 so that the plotting symbols will correspond to each month (1 = January, 2 = February, ..., 0 = October, A = November, and B = December).

```
MTB > SET 'AIRLINE.DAT' C1
C1
    112     118     132     129    .   .   .
```

```
MTB > NAME C1 'PASSNGRS'
MTB > TSPLOT 'PASSNGRS'
```

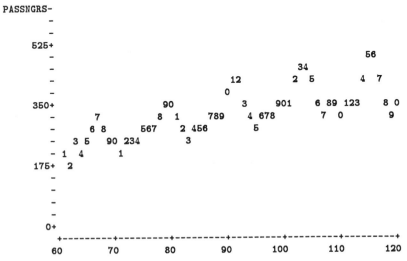

Only a portion of the plot is shown here. Notice how both the monthly patterns and the overall trend are displayed in this plot.

COMMAND REFERENCE

Listed below are the general forms of the Minitab commands discussed in this chapter.

TABLE C by C by C ...

ROWPERCENT - percentages based on row totals

COLPERCENT - percentages based on column totals

COUNT - frequency count

TOTPERCENTS - percentages of two-way table

FREQUENCIES in C

CHISQUARE - calculate chi-squared statistic

LAYOUT with first **K** variables for rows, next **K** for columns

The number of column variables must be between 0 and 2.

PLOT data in C versus data in C

LPLOT C vs C, labeled by **C**
CORRELATION of C,...,C
TSPLOT [with period **K**, starting at **K**] time series data in **C**

COMPUTER EXERCISES

5.1 Create a data set for the new car data shown in Table 5.59, including one categorical variable for body type and one for transmission type, plus a variable containing the frequencies of each combination. Be sure to describe your coding assignments for each variable. Use the TABLE command to compute the following percentages.

(a) The percentage of the 2-door cars that were sold with manual transmission.

(b) The percentage of the cars sold with automatic transmissions that had 4 doors.

(c) The percentage of all the cars sold that had both 4 doors and automatic transmission.

Compute the measure of association ϕ for this table. Do the number of doors and the type of transmission appear to be independent?

5.2 Create a data set with the information in Table 5.61 on foreign language proficiency versus language use after graduation.

(a) Tabulate the information in the same layout as Table 5.61.

(b) Summarize the information on proficiency into two groups by combining 'Average', 'Better than Average', and 'Extremely High' into the single category 'High', and produce a 2 × 2 table versus language use. [Hint: use the CODE command and create a new categorical variable. Minitab will add the frequencies across all rows with common values of the categories.]

5.3 Create a data set with the information in Table 5.64 on distributions of reading errors in three ability groups. Produce a table that shows the relative frequency of each type of error for each reading group. Compare the three frequency distributions.

5.4 Create a data set based on the information in Table 5.35 on classification of social scientists by age, productivity score, and party vote. Use the TABLE command to produce a table that includes the answer to each question below.

(a) Find the percentage of social scientists who voted Democratic in 1952.

(b) Among the social scientists in the 41-50 age group, what percentage are in the medium productivity group?

(c) Among those social scientists who did not vote Democratic in 1952, what percentage were 40 years or younger?

(d) Among those social scientists 51 years or older, what percentage had both a low productivity score and did not vote Democratic?

(e) Tabulate the data in the layout shown in Table 5.35.

5.5 Create a data set with the information on interviewers in Table 5.66, including numerical variables for competitiveness and productivity, and a categorical variable for the section.

(a) Obtain a scatter plot of productivity (on the y axis) versus competitiveness (on the x axis), with each point representing an interviewer in the A section plotted with an 'A' and each point representing one in the B section plotted with a 'B'. [Hint: this depends on having chosen your category codes correctly.] What appears to be true about the relationship between productivity and competitiveness in each section?

(b) Calculate the correlation coefficient of productivity with competitiveness for each section separately. [Hint: use the COPY command to select the appropriate rows.] Do the correlation coefficients show the same relationships as the plots?

5.6 For the data for the second grade pupils in Table 2.23 [IQ.DAT], do the following:

(a) Compute the gain in reading achievement for each child. Plot these against initial reading achievement. Calculate the correlation coefficient. Does there appear to be a relation?

(b) Plot final reading achievement score versus verbal IQ and initial reading achievement versus verbal IQ. Compute the correlation coefficient for each plot. Which appears to be more strongly related to verbal IQ, initial reading achievement or final reading achievement?

5.7 Read in the data on 24 students in a statistics course from Table 2.24 [STUDENT.DAT].

(a) Tabulate LOGIC versus SETS, showing counts and overall percentages. Compute ϕ. Does familiarity with logic appear to be associated with familiarity with set theory?

(b) Produce a table with LOGIC and SETS and sets across, and MAJOR down.

(c) Tabulate the number of semesters of statistics versus the number of semesters of analysis.

5.8 Read in the data on average speed and noise levels from Table 5.67 [AUTO.DAT]. Plot noise level, y, against average speed, x. Answer the questions in Exercise 5.22.

5.9 Read in and explore the 1973 weather data on 66 cities in [WEATHER-.DAT]. Use the techniques discussed in this chapter to examine possible relationships between the weather and location variables. Generate plots and tabulations to present your findings.

NOTES

1. The classification variables for the TABLE command must have integer values between -10000 and 10000 (between -1000 and 1000 in earlier versions of Minitab).

PART THREE

PROBABILITY

6 BASIC IDEAS OF PROBABILITY

INTRODUCTION

Though it is hard to get a computer to toss a real coin, it is possible to have the computer act as if it were tossing a coin or using another randomizing device. Minitab has a general command[1]

> **RANDOM K** trials, put in **C**

which generates random ("pseudo-random") numbers that we can interpret as if they were the results of physical randomizations. Subcommands are used to specify the particular random distribution.

6.3 PROBABILITY IN TERMS OF EQUALLY LIKELY CASES

The most common random experiment is the tossing of a fair coin. We can have Minitab perform an analagous randomization with RANDOM and the subcommand

> *RANDOM*
> **BERNOULLI** trials with $p = .5$

Minitab will generate a 1 or a 0 with equal probability. We can think of this as 1 representing a head and 0 representing a tail. For example

```
MTB > RANDOM 1 TOSS, PUT INTO C1;
SUBC> BERNOULLI WITH P=.5.
MTB > PRINT C1
C1
    1
```

The result is a 1 (head). If we do the same command over and over we will get some 1's and some 0's. If we want to see the results of several tosses without having to give the command multiple times, we use the command

> **RANDOM K** tosses, put into **C**

with the same subcommand. For instance

```
MTB > RANDOM 10 TOSSES, PUT INTO C1;
SUBC> BERNOULLI WITH P=.5.
MTB > PRINT C1
C1
    1    1    0    0    1    0    0    1    1    1
```

The results of these particular 10 coin flips are 6 heads and 4 tails, and the results of the coin flips are available in C1 for possible further use. If we give the same command again, we get a different sequence of outcomes.

Another common random event is rolling a 6-sided die. We can use the subcommand

```
RANDOM
    INTEGERS from 1 to 6
```

to obtain a random number from 1 to 6 with equal probabilities, the same as rolling a die. For example

```
MTB > RANDOM 1 ROLL, PUT IN C1;
SUBC> INTEGERS FROM 1 TO 6.
MTB > PRINT C1
C1
    4
```

gives a 4 as the result. As above, the results of rolling several dice can be obtained by the command

```
RANDOM K rolls, put into C
```

and the same subcommand. For instance

```
MTB > RANDOM 10 ROLLS, PUT IN C1;
SUBC> INTEGERS FROM 1 TO 6.
MTB > PRINT C1
C1
    6    3    5    5    1    3    4    2    3    4
```

gives the results of rolling 10 dice.

6.8   RANDOM SAMPLING; RANDOM NUMBERS

6.8.1   Random Numbers

Random Numbers With 2 Outcomes.   The subcommand

```
RANDOM
    BERNOULLI trials with p = K
```

generates random numbers 0 and 1 with the given probability of a 1 on each trial. We saw above how this could be used to obtain tosses from a fair coin. If we want to simulate an unfair coin, say with 60% probability of heads and 40% probability of tails, we do as follows:

```
MTB > RANDOM 10 TOSSES, PUT INTO C1;
SUBC> BERNOULLI WITH P=.6.
MTB > PRINT C1
C1
    1    0    1    1    0    0    1    1    1    1
```

Random Numbers With More Than 2 Outcomes. If we have random trials with more than two possible outcomes, each equally likely, we use the RANDOM command with the subcommand

---
RANDOM

INTEGERS between **K** and **K**

We saw above how this is used to obtain random die rolls. We can also use this command to produce random digits from 0 to 9:[2]

```
MTB > RANDOM 100 DIGITS, PUT IN C1;
SUBC> INTEGERS FROM 0 TO 9.
MTB > PRINT C1
C1
    3    4    8    0    9    0    5    7    9    8    2    4    9    9
    1    4    4    4    0    1    6    8    0    0    4    2    5    2
    7    5    8    7    9    7    3    3    5    1    5    2    5    3
    5    0    3    1    6    2    8    1    8    4    4    1    6    9
    3    4    7    8    1    0    8    0    3    6    9    5    0    3
    0    4    8    2    9    1    4    4    6    8    7    6    8
    8    9    7    3    9    8    5    4    5    6    7    1    4    1
    5    9
```

After generating a set of simulated trials, we can use other Minitab commands to examine the results. For instance, continuing the example above, we can examine the frequency with which each digit occurred:

```
MTB > HISTOGRAM C1

Histogram of C1   N = 100

Midpoint   Count
       0      10   **********
       1      10   **********
       2       6   ******
       3       9   *********
       4      15   ***************
       5      11   ***********
       6       7   *******
       7       8   ********
       8      13   *************
       9      11   ***********
```

Random Numbers with Unequal Probabilities. If we want to simulate a process with several outcomes with unequal probabilities, we can use the subcommand

> *RANDOM*
> **DISCRETE** distribution, values in C, probabilities in C

The codes for the possible outcomes and the probability of each must be read into Minitab before generating the trials. For instance, suppose we want to simulate a spinner such as is shown in Figure 6.1, where the three spaces on the spinner account for 40%, 50%, and 10% of the circle, respectively. Then we can code the three colors as 1, 2, and 3 and read them in with their probabilities to generate random spins of the spinner, as follows:

```
MTB > READ C1,C2
DATA> 1 0.4
DATA> 2 0.5
DATA> 3 0.1
DATA> END
     3 ROWS READ
MTB > RANDOM 100 DRAWS, PUT IN C3;
SUBC> DISCRETE VALUES IN C1, PROBABILITIES IN C2.
MTB > HISTOGRAM C3

Histogram of C3   N = 100

Midpoint   Count
       1      41   *****************************************
       2      45   *********************************************
       3      14   **************
```

Note that the number of entries in the column of results will usually be different from the number of entries in the first two columns.

6.8.2   Sampling from a Finite Population

Sometimes we want to sample without replacement from a finite population. The Minitab command to do this is

> **SAMPLE K** rows from C, put into C (without replacement)

For instance, if are working with the income data from 49 families given in Table 2.17, and we want to choose 10 families to receive a follow-up questionaire, we proceed as follows:

```
MTB > READ 'INCOME.DAT' INTO C1, C2
      49 ROWS READ
  ROW      C1      C2

    1    31031      85
    2    31059     135
    3    31069     200
    4    31159      50
      .    .    .

MTB > NAME C1 'ID'
MTB > NAME C2 'INCOME'
MTB > SAMPLE 10 FROM 'ID' , PUT IN C3
MTB > PRINT C3
C3
   36164    31595    36232    31069    36796    32247    36980    31633
   32561    31951
```

These are the respondent identification numbers of the families to be sent the follow-up questionaire.

If in doing our sampling we want to keep all of the variables together, we use the extended form of the SAMPLE command

---
SAMPLE K rows from **C,...,C**, put into **C,...,C**

The same rows are taken from all columns. For instance, with the income data we have

```
MTB > SAMPLE 10 FROM 'ID','INCOME', PUT IN C3, C4
MTB > PRINT C3,C4
  ROW      C3      C4

    1    32273     120
    2    31209      70
    3    31785     150
    4    31059     135
    5    31469      65
    6    32317     110
    7    31867     147
    8    31927      84
    9    36514      72
   10    36980      95
```

If we want to sample without replacement from a set of integers, we use the command

---
GENERATE the integers from **K** to **K**, put in **C**
---

to obtain the column to sample from, as shown:

```
MTB > GENERATE 1 100 C1
MTB > SAMPLE 10 C1 C2
MTB > PRINT C2
C2
    2    84    15    33    69    17    53    73    66    76
```

COMMAND REFERENCE

Listed below are the general forms of the Minitab commands discussed in this chapter.

RANDOM K observations into each of **C**,...,**C**
    BERNOULLI trials with $p = $ **K**
    INTEGERS uniform on $a = $ **K** to $b = $ **K**
    DISCRETE distribution with x values in **C**, probabilities in **C**
SAMPLE K rows (without replacement) from **C**,...,**C**, put into **C**,...,**C**
GENERATE the integers from **K** to **K**, put in **C**

COMPUTER EXERCISES

6.1 Generate 25 simulated fair coin tosses into column C1, and another 25 tosses into C2. This simulates 25 pairs of coin tosses. Generate tables of the observed frequencies and observed proportions of outcomes as shown in Exercise 6.1.

6.2 Pick a number from 1 to 10. Write it down. Now have the computer generate a random number from 1 to 10. Did you pick the same number? Repeat this for a total of 10 trials. How many times did you agree? How many times would you expect to agree by chance alone?

6.3 Write down a coding list for a deck of cards into the numbers 1 – 52. Use Minitab to generate a random bridge hand of 13 cards from a full deck. [Hint: You can only receive one of each card in your hand.]

NOTES

1. Earlier versions of Minitab had separate commands for random trials from each distribution, as follows:

BTRIALS K Bernoulli trials with $p = $ **K**, put into **C**

IRANDOM K random integers between **K** and **K**, put into **C**

DRANDOM K obs., values in **C**, probabilities in **C**, put into **C**

---

2. In earlier versions of Minitab, the results of random trials are automatically printed out. If we are generating random numbers but do not want to see the results printed out, the command

NOPRINT

turns off printing of this output for all subsequent random number commands. The NOPRINT command also turns off the list of values printed by commands such as READ and SET when reading data from a file. The command

PRINT

turns this printing back on. Give the NOPRINT command before generating the random numbers, and use PRINT to turn printing back on if you want to see the results for later random number commands.

7 SAMPLING DISTRIBUTIONS

INTRODUCTION

In addition to the ability to generate discrete random numbers as seen in Chapter 6, the Minitab RANDOM command has subcommands to generate continuous random variables from a number of common probability distributions.[1] Minitab also has functions to compute probability distribution, cumulative distribution, and inverse distribution functions for many common probability distributions.

7.1 PROBABILITY DISTRIBUTIONS

7.1.4 Uniform Distributions

The Discrete Uniform Distributions. The discrete uniform distribution can be generated by rescaling the output from the RANDOM command

> **RANDOM K** trials, put data into **C**

with the INTEGERS subcommand

> *RANDOM*
> **INTEGERS** from **K** to **K**

For instance, to generate the discrete uniform distribution on $0.00 - 0.99$ we take random integers from $0 - 99$ and divide the results by 100:

```
MTB > RANDOM 10 VALUES, PUT IN C1;
SUBC> INTEGERS FROM 0 TO 99.
MTB > PRINT C1
C1
      8     10     27     69     66     35     63     62      4     12

MTB > DIVIDE C1 BY 100, PUT IN C1
MTB > PRINT C1
C1
   0.08   0.10   0.27   0.69   0.66   0.35   0.63   0.62   0.04   0.12
```

The Continuous Uniform Distributions. Simulated samples from a continuous uniform distribution on the interval from 0 to 1 are generated by the subcommand

---
RANDOM
> **UNIFORM** distribution [on **K** to **K**]

If limits are not given, a uniform on (0,1) is assumed. For instance

```
MTB > RANDOM 10 VALUES, PUT IN C2;
SUBC> UNIFORM ON 0 TO 1.
MTB > PRINT C2
C2
   0.272817    0.462704    0.675440    0.368925    0.511900    0.983092
   0.760007    0.277417    0.309619    0.592996
```

Here Minitab is printing out only 6 significant figures of each number. In the Minitab worksheet these numbers are represented to greater than 6 digits of accuracy, and so can be thought of as representing a continuous distribution. Samples with uniform distributions on other intervals can be obtained by other parameters to the UNIFORM subcommand.

7.2 THE FAMILY OF NORMAL DISTRIBUTIONS

Probability distribution functions are computed in Minitab with the command

---
> **PDF** for values in **E**, [put results in **E**]
---

and a subcommand to specify the distributional family and the parameters of the distribution. For the normal distribution, the subcommand is

---
PDF
> **NORMAL** [with $\mu = $ **K**, $\sigma = $ **K**]

A standard normal distribution ($\mu = 0$, $\sigma = 1$) is assumed if the parameters are not specified. For instance, the probability density of the standard normal distribution at 1.0 is calculated by

```
MTB > PDF AT 1.0;
SUBC> NORMAL MU=0, SIGMA=1.
   1.00     0.2420
```

The probability density at 1.0 is 0.242. The PDF command can operate on a column of values to generate a plot of the normal distribution, as follows:[2]

```
MTB > SET C1
DATA> -4:4/.2
DATA> END
MTB > PDF AT C1, PUT IN C2;
SUBC> NORMAL MU=0, SIGMA=1.
MTB > PLOT C2 VS C1
```

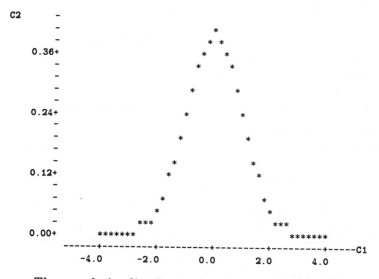

The cumulative distribution function can be computed with the command

CDF for values in **E** [put results in **E**]

and the same subcommands as PDF to specify the distribution. For instance the value of the cumulative standard distribution at 1.96 can be computed by

```
MTB > CDF AT 1.96;
SUBC> NORMAL MU=0, SIGMA=1.
     1.96    0.9750
```

Thus the probability that a standard normal is less than or equal to 1.96 is 0.975. To reverse the process, and determine the point where the cumulative distribution equals a given probability, we calculate the inverse cumulative distribution function with the command

INVCDF for values in **E** [put results in **E**]

and the same subcommands as PDF and CDF to specify the distribution. For instance, to reverse the calculation above, we have

```
MTB > INVCDF AT .975;
SUBC> NORMAL MU=0, SIGMA=1.
     0.98    1.9600
```

These commands can be used in place of tables of probabilities and percentage points in many cases.

Simulated samples from a normal distribution are generated by the RANDOM command with the subcommand

---

RANDOM

NORMAL [with $\mu = $ **K**, $\sigma = $ **K**]

Values from a standard normal distribution are obtained by using $\mu = 0$ and $\sigma = 1$ (which are the default if not specified). For example

```
MTB > RANDOM 5 VALUES, PUT IN C1;
SUBC> NORMAL MU=0, SIGMA=1.
MTB > PRINT C1
C1
 -0.87313    0.14418    1.03686   -0.32576    1.13174
```

Simulated normal samples from other normal distributions can be obtained by specifying the desired μ and σ. For instance, to generate 5 observations from a normal distribution with $\mu = 3$ and $\sigma = 2$ we do as follows:

```
MTB > RANDOM 5 VALUES, PUT IN C1;
SUBC> NORMAL MU=3, SIGMA=2.
MTB > PRINT C1
C1
   2.43312    2.94208    4.62190   -0.09072   -1.68087
```

7.3 SAMPLING FROM A POPULATION

We can examine simulated samples with the commands we have already learned such as the HISTOGRAM command

---

HISTOGRAM of C

---

in order to understand how real samples vary. For instance, the histogram of a simulated normal sample will resemble the normal bell curve but will vary somewhat according to the sample:

```
MTB > RANDOM 100 VALUES, PUT IN C1;
SUBC> NORMAL MU=5, SIGMA=10.
MTB > HISTOGRAM C1

Histogram of C1   N = 100
```

```
Midpoint   Count
   -20       1   *
   -15       6   ******
   -10      11   ***********
    -5      15   ***************
     0      13   *************
     5      17   *****************
    10      21   *********************
    15       8   ********
    20       5   *****
    25       3   ***
```

7.4   SAMPLING DISTRIBUTION OF A SUM

We can experiment with sums of random variables and obtain sample histograms. For instance, to examine a sample distribution of the sums of pairs of dice we generate two columns of dice results and add them:

```
MTB > RANDOM 100 VALUES, PUT IN C1, C2;
SUBC> INTEGERS FROM 1 TO 6.
MTB > LET C3 = C1 + C2
MTB > HISTOGRAM C3

Histogram of C3   N = 100

Midpoint   Count
     2       4   ****
     3       7   *******
     4       2   **
     5      11   ***********
     6      12   ************
     7      24   ************************
     8       9   *********
     9      15   ***************
    10       5   *****
    11       9   *********
    12       2   **
```

(Note that several columns of random numbers can be generated with one RANDOM command.) Adding a third column of dice results we obtain a sample distribution of the sums of three dice:

```
MTB > RANDOM 100 VALUES, PUT IN C4;
SUBC> INTEGERS FROM 1 TO 6.
MTB > LET C5 = C3 + C4
MTB > HISTOGRAM C5
```

```
Histogram of C5   N = 100

Midpoint   Count
       4       6  ******
       6       5  *****
       8      17  *****************
      10      24  ************************
      12      18  ******************
      14      17  *****************
      16      12  ************
      18       1  *
```

7.5   THE BINOMIAL DISTRIBUTION

We saw in Chapter 6 how to generate Bernoulli trials with outcomes 0 and 1 with the RANDOM command with subcommand

---
RANDOM
   BERNOULLI trials with $p = $ **K**

The number of 1's in n Bernoulli trials has a binomial distribution. Binomial random values can be generated directly with the RANDOM subcommand

---
RANDOM
   BINOMIAL with $n = $ **K**, $p = $ **K**

To simulate the results of tossing 5 fair coins and counting the number of heads we do

```
MTB > RANDOM 1 TRIAL, PUT IN C1;
SUBC> BINOMIAL WITH N=5, P=.5.
MTB > PRINT C1
C1
    2
```

Among the 5 coins there were 2 heads. To simulate 100 repetitions of tossing 5 fair coins and counting the number of heads we do as follows

```
MTB > RANDOM 100 TRIALS, PUT IN C1;
SUBC> BINOMIAL WITH N=5, P=.5.
MTB > HISTOGRAM OF C1
```

```
Histogram of C1   N = 100

Midpoint   Count
       0       1  *
       1      18  ******************
       2      30  ******************************
       3      33  *********************************
       4      15  ***************
       5       3  ***
MTB > MEAN OF C1
   MEAN     =      2.5200
MTB > LET K1 = STDEV(C1)**2
MTB > PRINT K1
K1       1.16121
```

The mean of 2.52 and variance of 1.16 computed for this sample are close to the population mean and variance of 2.5 and 1.25 as computed by the formulas in Section 7.5.3.

Binomial trials with other values of n and p are generated in the same way. For instance, to obtain samples of the numbers of heads from three tosses of an unfair coin with a 40% chance of heads on each toss, we have:

```
MTB > RANDOM 100 TRIALS, PUT IN C1;
SUBC> BINOMIAL WITH N=3, P=.4.
MTB > HISTOGRAM OF C1

Histogram of C1   N = 100

Midpoint   Count
       0      22  *********************
       1      39  **************************************
       2      31  *******************************
       3       8  ********
MTB > MEAN OF C1
   MEAN     =      1.2500
MTB > LET K1 = STDEV(C1)**2
MTB > PRINT K1
K1       0.795455
```

The mean of 1.25 and variance of 0.80 are fairly close to the population mean and variance for this distribution of 1.2 and 0.72.

It is often useful to have tables of the probabilities of the number of heads in n trials for a given p. The command[3]

CDF - cumulative distribution
---

with no arguments and the the subcommand

CDF

BINOMIAL table for $n = $ **K**, $p = $ **K**
---

prints out the entire cumulative distribution function for a binomial distribution with the given parameters. For instance, corresponding to our example above we have

```
MTB > CDF;
SUBC> BINOMIAL FOR N=3, P=.4.

    BINOMIAL WITH N =   3  P = 0.400000
      K  P( X LESS OR = K)
      0           0.2160
      1           0.6480
      2           0.9360
      3           1.0000
```

The probabilities can be stored in a column for further use by using the form

---
CDF for values in **C**, put in **C**
---

with the first column containing the integers from 1 to n.

7.6 THE LAW OF AVERAGES (LAW OF LARGE NUMBERS)

Using the random values generated by Minitab we can illustrate the Law of Averages. To do so we shall generate a sample of observations from a uniform distribution and take the average. We shall then repeat the procedure several times to see how the sample average varies around the true value.

In order to make repeating the commands easier, we shall have Minitab remember a series of commands and repeat them for us. This is done by the command

---
STORE commands

After the STORE command, Minitab switches to a **STOR>** prompt and does not execute the commands which we then enter. When we have entered the set of commands, we give the command

---
END storing commands
---

to return to the usual Minitab mode. After storing some commands, we can execute them several times with the command

---
EXECUTE stored commands **K** times

Here we store the commands to generate 6 uniform random variables and calculate their mean. We execute these commands 10 times, and then type the 10 mean values back into Minitab so that we can see how they vary around the true value.

```
MTB > RANDOM 6 VALUES, PUT IN C1;
MTB > UNIFORM ON 0 TO 1.
MTB > MEAN OF C1
MTB > END OF STORED COMMANDS
MTB > EXEC 10 TIMES
```

[from this point, MINITAB both types and executes the commands]

```
MTB > RANDOM 6 VALUES, PUT IN C1;
SUBC> UNIFORM ON 0 TO 1.
MTB > MEAN OF C1
   MEAN     =      0.69076
MTB > END OF STORED COMMANDS
MTB > RANDOM 6 VALUES, PUT IN C1;
SUBC> UNIFORM ON 0 TO 1.
MTB > MEAN OF C1
   MEAN     =      0.50325
MTB > END OF STORED COMMANDS
MTB > RANDOM 6 VALUES, PUT IN C1;
SUBC> UNIFORM ON 0 TO 1.
MTB > MEAN OF C1
   MEAN     =      0.61470
MTB > END OF STORED COMMANDS
MTB > RANDOM 6 VALUES, PUT IN C1;
SUBC> UNIFORM ON 0 TO 1.
MTB > MEAN OF C1
   MEAN     =      0.51669
MTB > END OF STORED COMMANDS
MTB > RANDOM 6 VALUES, PUT IN C1;
SUBC> UNIFORM ON 0 TO 1.
MTB > MEAN OF C1
   MEAN     =      0.57193
MTB > END OF STORED COMMANDS
MTB > RANDOM 6 VALUES, PUT IN C1;
SUBC> UNIFORM ON 0 TO 1.
MTB > MEAN OF C1
   MEAN     =      0.48258
MTB > END OF STORED COMMANDS
MTB > RANDOM 6 VALUES, PUT IN C1;
SUBC> UNIFORM ON 0 TO 1.
MTB > MEAN OF C1
   MEAN     =      0.65745
MTB > END OF STORED COMMANDS
MTB > RANDOM 6 VALUES, PUT IN C1;
SUBC> UNIFORM ON 0 TO 1.
MTB > MEAN OF C1
   MEAN     =      0.73696
MTB > END OF STORED COMMANDS
MTB > RANDOM 6 VALUES, PUT IN C1;
```

```
SUBC> UNIFORM ON 0 TO 1.
MTB > MEAN OF C1
    MEAN     =      0.51967
MTB > END OF STORED COMMANDS
MTB > RANDOM 6 VALUES, PUT IN C1;
SUBC> UNIFORM ON 0 TO 1.
MTB > MEAN OF C1
    MEAN     =      0.38527
MTB > END OF STORED COMMANDS

[MINITAB returns control to the user at this point]
```

Now to analyze the mean values, we type them back into Minitab:

```
MTB > SET C2
DATA> 0.69076,0.50325,0.61470,0.51669,0.57193
DATA> 0.48258,0.65745,0.73696,0.51967,0.38527
DATA> END
MTB > HISTOGRAM OF C2 , FIRST MIDPOINT 0.1, INTERVAL WIDTH 0.1

Histogram of C2   N = 10

Midpoint   Count
  0.100      0
  0.200      0
  0.300      0
  0.400      1   *
  0.500      4   ****
  0.600      2   **
  0.700      3   ***

MTB > DESCRIBE C2

               N      MEAN    MEDIAN    TRMEAN     STDEV    SEMEAN
C2            10    0.5679    0.5458    0.5696    0.1073    0.0339

             MIN       MAX        Q1        Q3
C2        0.3853    0.7370    0.4981    0.6658
```

For averages of 6 observations the means are fairly tightly grouped around the population mean value of 0.50.

Next we shall examine the distribution of the mean of 25 uniform random variables. Doing this 200 times will give us a fairly good picture of the distribution of the means. In order to avoid having to write down 200 means and type them back into Minitab we shall store a slightly more complicated set of commands.

```
MTB > STORE
MTB > LET K1=K1+1
MTB > RANDOM 25 VALUES, PUT IN C1;
MTB > UNIFORM.
MTB > LET C2(K1) = MEAN(C1)
MTB > END
```

We shall initialize the process with K1 = 0, and then LET K1 = K1+1 in each execution of the stored commands, which means that K1 will be 1 on the first execution of the stored commands, 2 on the second, etc. In the LET statement, a column followed by a row in parentheses can be assigned a value, putting the value into the particular row of the column. Here

```
MTB > LET C2(K1) = MEAN(C1)
```

stores the mean of C1 in row K1 of C2.[4] After Minitab executes the stored commands 200 times, C2 will contain the 200 means, one in each row.[5]

```
MTB > NOTE SET UP STARTING VALUES
MTB > LET K1=0
MTB > EXECUTE 200 TIMES

[MINITAB takes over typing from this point on]

MTB > LET K1=K1+1
MTB > RANDOM 25 VALUES, PUT IN C1;
SUBC> UNIFORM.
MTB > LET C2(K1) = MEAN(C1)
MTB > END

[198 repetitions deleted from this display]

MTB > LET K1=K1+1
MTB > RANDOM 25 VALUES, PUT IN C1;
SUBC> UNIFORM.
MTB > LET C2(K1) = MEAN(C1)
MTB > END

[We resume typing at this point]

MTB > HISTOGRAM OF C2;
SUBC> START AT 0.1 END AT 0.9;
SUBC> INCREMENT 0.1.

Histogram of C2   N = 200
Each * represents 5 obs.

Midpoint    Count
   0.100       0
   0.200       0
   0.300       2   *
   0.400      44   *********
   0.500     117   ***********************
   0.600      37   ********

MTB > MEAN OF C2
   MEAN    =      0.49243
MTB > STANDARD DEVIATION OF C2
   ST.DEV. =      0.062584
```

The means of 25 uniform random variables are very tightly distributed around the population mean of 0.50.

7.7   THE NORMAL DISTRIBUTION OF SAMPLE MEANS (CENTRAL LIMIT THEOREM)

We can demonstrate the Central Limit Theorem using Minitab. In fact the simulated data generated above for means of 25 uniform random variables is useful here as well. If we allow the HISTOGRAM function to choose the scale of the histogram, we can see the shape of the distribution of means, as follows

```
MTB > HISTOGRAM OF C2

Histogram of C2   N = 200

Midpoint   Count
   0.34       2   **
   0.36       3   ***
   0.38       8   ********
   0.40       9   *********
   0.42      10   **********
   0.44      14   **************
   0.46      28   ****************************
   0.48      26   **************************
   0.50      23   ***********************
   0.52      21   *********************
   0.54      19   *******************
   0.56      15   ***************
   0.58       9   *********
   0.60       8   ********
   0.62       5   *****
```

From the histogram we can see that these means are distributed quite normally around the true value of 0.50, similarly to the results shown in Figures 7.19 .

COMMAND REFERENCE

Here are the complete versions of commands discussed in this chapter:

RANDOM data into **C**
PDF for values in **E**, [put results in **E**]
CDF for values in **E** [put results in **E**]
INVCDF for values in **E** [put results in **E**]
      RANDOM, PDF, CDF, and INVCDF all have the subcommands
  INTEGERS from **K** to **K**
  UNIFORM distribution [on **K** to **K**]

NORMAL [with $\mu = $ **K**, $\sigma = $ **K**]
BERNOULLI trials with $p = $ **K**
BINOMIAL with $n = $ **K**, $p = $ **K**
HISTOGRAM of C
STORE commands [in **'FILENAME'**]
END storing commands
EXECUTE stored commands [from **'FILENAME'**] **K** times

COMPUTER EXERCISES

7.1 Generate 10 continuous uniform values from a uniform distribution on the interval from 2 to 4.

7.2 Plot the cumulative distribution function of a normal distribution with mean 3 and standard deviation 2. [Hint: the interval from -5 to 11 would be a good choice for a plotting region. Why?]

7.3 Generate 10 normal random values from a normal distribution with a mean of 100 and a standard deviation of 10. Generate the histogram of the 10 values. Do this 5 times. Do the same for sets of 100 normal random values.

7.4 Generate the binomial table for $n = 5$, $p = .5$, and store the probabilities in a column. Calculate the population mean and variance of this binomial distribution from these probabilities using Minitab. Do your answers agree with the theoretical formulas given in section 7.5.3?

7.5 Try out the Law of Averages for different sample sizes:

(a) Generate 10 uniform random values and take their average. Repeat this for a total of 20 times. Tabulate the 20 averages, and generate a histogram, the mean, and the standard deviation of the 20 values. Use interval widths of 0.1 for the histogram.

(b) Do the same for 20 sets of 1000 uniform random values. Compare with the set of results above and the set of results in this chapter. Does the Law of Averages seem to be working?

NOTES

1. In previous versions of Minitab, separate commands were used for obtaining random numbers for each distribution:

IRANDOM K random integers from **K** to **K**, put into **C**

URANDOM K uniform observations, put into **C**

NRANDOM K normal observations with $\mu = $ **K**, $\sigma = $ **K**, put into **C**

BTRIALS K Bernoulli trials with $p = $ **K**, put in **C**

BRANDOM K binomial experiments with $n = $ **K**, $p = $ **K**, put into **C**

---

2. A special patterned data input feature of the SET command is used to generate the x-axis values in the example. A pattern of the form $a : b/c$ indicates a list of evenly spaced values from a to b with an interval of c (an interval of 1 is the default if the $/c$ is omitted). A list of values can be repeated by placing it in parentheses and placing a repeat factor immediately before or after the parentheses. A number before the list repeats the entire list the indicated number of times. A number after the list repeats each value individually the indicated number of times. For instance 3(1:2,5) indicates the values 1 2 5 1 2 5 1 2 5, and (1:2,5)3 indicates the values 1 1 1 2 2 2 5 5 5.

3. In earlier versions of Minitab, the binomial tables are obtained with the command

---

BINOMIAL table for $n = $ **K**, $p = $ **K**, [store probabilities in **C**]

---

4. In earlier versions of Minitab, the command

SUBSTITUTE K into row **K** of **C**

---

would be used to store the mean for a trial (in K2) into row K1 of C2.

5. In earlier versions of Minitab, it may be necessary to set up C2 as a column of length 200 before starting. One possibility is to use the GENERATE command

---

GENERATE integers from **K** to **K**, put in **C**

---

to create C2 as a column of length 200 for storing the results.

PART FOUR

STATISTICAL INFERENCE

8 USING A SAMPLE TO ESTIMATE POPULATION CHARACTERISTICS

INTRODUCTION

Many of the Minitab commands we have already learned can be used for calculation of estimates of population characteristics from a sample. In addition to providing point estimates, Minitab can calculate confidence intervals for estimates of mean values.

8.1 ESTIMATION OF A MEAN BY A SINGLE NUMBER

If the data in a column C consist of the set of observations from a sample, $x_1, \ldots, x_n$, then the command

MEAN of the values in **C** [put into **K**]

calculates the sample mean

$$\overline{x} = \frac{1}{n} \sum_{i=1}^{n} x_i.$$

The command AVERAGE does exactly the same thing as MEAN.

Suppose the 23 pupils in Table 2.23 on IQ and reading achievement [IQ.DAT] are a random sample of all second-grade pupils, and we are interested in learning about the verbal IQ (in C3, named VERBALIQ) of second-grade pupils. It is always a good idea to look at the data:

```
MTB > HIST OF 'VERBALIQ'

 Histogram of VERBALIQ    N = 23
```

```
Midpoint    Count
      80      2    **
      85      3    ***
      90      2    **
      95      5    *****
     100      4    ****
     105      3    ***
     110      2    **
     115      0
     120      2    **
```

We can see that in our sample, verbal IQ ranges from about 80 to about 120, centered approximately around 100, and roughly the same should be true of the population. To estimate the mean verbal IQ of all second-grade pupils we do

```
MTB > MEAN OF 'VERBALIQ'
   MEAN    =       97.565
```

This gives 97.6 as the mean verbal IQ of this sample of second-grade pupils, and, assuming the class is a representative sample, an estimate of the mean verbal IQ for the population of all second-grade pupils.

8.2 ESTIMATION OF VARIANCE AND STANDARD DEVIATION

8.2.1 Standard Deviation

In the same fashion as above, the command

STDEV of **C**, [put standard deviation into **K**]

(or the equivalent command STANDARD) calculates the sample standard deviation

$$s = \sqrt{\frac{\sum_{i=1}^{n}(x_i - \overline{x})^2}{n-1}}.$$

Note that because in most cases we are working with sample data, rather than the whole population, Minitab does the calculation using $n-1$ in the denominator.

8.2.2 Variance

Minitab does not provide a command to directly calculate the sample variance, but we can easily obtain it, by storing the standard deviation in a constant and squaring it using the LET command, as shown below.

```
MTB > STDEV OF 'VERBALIQ', RESULT IN K1
   ST.DEV. =       10.782
MTB > LET K2 = K1**2
MTB > PRINT K2 (THE VARIANCE OF VERBAL IQ)
K2       116.257
```

This gives a sample standard deviation of 10.8 and a sample variance of 116.3 for the pupils in the class, which are estimates of the corresponding population values.

8.2.3   Standard Error of the Mean

The standard error of the mean of a sample can be calculated by $s/\sqrt{n}$. For the mean verbal IQ, we have

```
MTB > LET K3 = STDEV('VERBALIQ')/SQRT(COUNT('VERBALIQ'))
MTB > PRINT K3 (STANDARD ERROR OF THE MEAN)
K3       2.24825
```

The estimated mean verbal IQ of second-grade pupils is then 97.6 with a standard error of 2.2.

8.3   AN INTERVAL OF PLAUSIBLE VALUES FOR A MEAN

8.3.1   Confidence Intervals When the Standard Deviation Is Known

Minitab makes it easy to calculate confidence intervals for the population mean. If the population standard deviation σ is known and we desire a confidence level of **K** percent, then the command

ZINTERVAL [with **K**% confidence] assuming $\sigma =$ **K**, for data in **C**

will calculate the confidence interval

$$\left(\overline{x} - z\frac{\sigma}{\sqrt{n}}, \overline{x} + z\frac{\sigma}{\sqrt{n}} \right),$$

where z is the appropriate multiple of the standard deviation to achieve K percent confidence of covering the population mean, using tables of the normal distribution such as those given in Appendix II of the text.[1] If no confidence level is specified in the command, a 95% level is used.

Now, suppose we want to calculate a 95% confidence interval for the mean verbal IQ of second-grade pupils. If verbal IQs are assumed to be normally distributed, and we have reliable information that the standard deviation of verbal IQs in second-grade pupils is 10 beats/minute, then we do as follows:

```
MTB > ZINTERVAL WITH 95 PCT CONF., ASSUMING SIGMA = 10, FOR 'VERBALIQ'

THE ASSUMED SIGMA =10.0

                 N      MEAN    STDEV   SE MEAN   95.0 PERCENT C.I.
   VERBALIQ     23     97.57    10.78     2.09   (  93.47,  101.66)
```

So we are 95% confident that the population mean verbal IQ of second-grade pupils lies between 93.5 and 101.7. Note that we are reassured as to the correctness of our prior information on the standard deviation, by noting that the sample standard deviation of 10.8 is close to our assumed standard deviation of 10.

8.3.2 Confidence Intervals When the Standard Deviation Is Estimated

If, as is more often the case, we need to estimate both the mean and standard deviation from the data, the appropriate confidence interval for the population mean can be calculated by

TINTERVAL [with **K%** confidence] for data in **C**

which gives the interval

$$\left(\overline{x} - t_{n-1}\frac{s}{\sqrt{n}}, \overline{x} + t_{n-1}\frac{s}{\sqrt{n}} \right),$$

where t_{n-1} is the appropriate multiple of the standard deviation to achieve **K** percent confidence of covering the population mean, using tables of the Student's t-distribution with $n-1$ degrees of freedom, such as those given in Appendix III.[2] Note that this command has a slightly different form from ZINTERVAL, since we do not need to specify the value of σ. If the confidence level is not specified, a 95% confidence interval is calculated.

For the IQ data, we have

```
MTB > TINTERVAL WITH 95 PERCENT CONFIDENCE FOR 'VERBALIQ'

                 N      MEAN    STDEV   SE MEAN   95.0 PERCENT C.I.
   VERBALIQ     23     97.57    10.78     2.25   (  92.90,  102.23)
```

So based on this analysis, we can say we are 95% confident that the population mean verbal IQ lies between 92.9 and 102.2.

The subcommand

RANDOM, CDF, PDF, INVCDF
T with **K** degrees of freedom

to the RANDOM, CDF, PDF, and INVCDF commands provides random observations and distributional calculations for the t-distribution.

8.4   ESTIMATION OF A PROPORTION

8.4.1   Point Estimation of a Proportion

Minitab does not provide a function specifically for dealing with proportions, but it is easy to calculate the desired values. The most convenient data coding for calculating proportions is to let a 1 represent a positive outcome and a 0 represent a negative outcome as described in Section 3.3.5. If the data are in this form, then the average of these 1's and 0's is the proportion of 1's. For this kind of data the MEAN command can be used to calculate proportions. Categorical data that has been coded in a different format can be recoded using the command[3]

CODE (K,...,K) to K (K,...,K) to K ... in C put into C

This command changes all of the parenthesized values in the column to a new value. A range of values to be changed can be indicated by a colon, e.g. 1:5 indicates all values from 1 to 5.

For instance, suppose the data from Table 2.1 on occupations of heads of households form a random sample of households, and are coded as in Chapter 2 (professional = 1, sales = 2, clerical = 3, laborer = 4), and we are interested in estimating the population proportion of professionals. First we need to recode the data, then take the mean:

```
MTB > NOTE FIRST WE RECODE TO GET A 0,1 VARIABLE
MTB > CODE (2:4) TO 0 IN 'OCCUP', PUT IN C2
MTB > NAME C2 'PROFESS'
MTB > MEAN 'PROFESS' (THE PROPORTION OF PROFESSIONALS)
   MEAN    =      0.30000
```

Therefore 30% of the heads of households are professionals. Note that here we used the fact that the code for professionals was 1. For the other occupations, a second list would be used in CODE, as follows:

```
MTB > CODE (1,3:4) TO 0, (2) TO 1 IN 'OCCUP', PUT IN C3
MTB > NAME C3 'SALES'
MTB > MEAN 'SALES' (THE PROPORTION OF SALES OCCUPATIONS)
   MEAN    =      0.40000
```

8.4.2   Interval Estimation of a Proportion

If we store the results from the MEAN command in a constant, then all the other calculations to obtain a confidence interval for p can be done with the LET command. If the sample size is large, we can calculate the interval

$$\left(\hat{p} - z\sqrt{\frac{\hat{p}\hat{q}}{n}}, \hat{p} + z\sqrt{\frac{\hat{p}\hat{q}}{n}}\right),$$

where z is the appropriate normal value. One way to do this in Minitab is to precompute the estimated standard deviation $\sqrt{\hat{p}\hat{q}}$ and give this as the 'known' value of σ to the ZINTERVAL command. Continuing our previous example, we have

```
MTB > NOTE FOR THE POPULATION PROPORTION OF PROFESSIONALS
MTB > NOTE FIRST WITHOUT THE CONTINUITY CORRECTION
MTB > MEAN 'PROFESS' , RESULT IN K1
   MEAN    =     0.30000
MTB > LET K2 = 1 - K1
MTB > LET K4 = SQRT(K1*K2)
MTB > ZINTERVAL WITH 95 PERCENT CONFIDENCE,SIGMA= K4, ON 'PROFESS'

THE ASSUMED SIGMA =0.458

              N      MEAN    STDEV   SE MEAN   95.0 PERCENT C.I.
PROFESS      20     0.300    0.470    0.102  (   0.099,    0.501)
```

A 95% confidence interval for the population proportion of professionals among heads of households by this method is $(9.8\%, 50.1\%)$. In most cases, since the computer is doing all the work, we can use the continuity correction, as described in Appendix 8A, and calculate the interval

$$\left(\hat{p} - \frac{1}{2n} - z\sqrt{\frac{\hat{p}\hat{q}}{n}}, \hat{p} + \frac{1}{2n} + z\sqrt{\frac{\hat{p}\hat{q}}{n}} \right),$$

as is shown below:

```
MTB > NOTE NOW WITH THE CONTINUITY CORRECTION
MTB > COUNT 'PROFESS' K3 (SAMPLE SIZE)
   COUNT   =      20.000
MTB > LET K5 = 1/(2*K3) + 1.96*SQRT(K1*K2/K3)
MTB > LET K6 = K1 - K5
MTB > LET K7 = K1 + K5
MTB > PRINT K6,K7 (CONFIDENCE LIMITS)
K6      0.0741598
K7      0.525840
```

This gives the slightly larger but more accurate interval $(7.4\%, 52.6\%)$ for the proportion of professionals.

8.5 ESTIMATION OF A MEDIAN OR OTHER PERCENTILE

8.5.1 Point Estimation of a Percentile

The Median. The command

MEDIAN of the values in **C** [put into **K**]

computes the median of a sample.

For instance, if we are interested in the population median of bulb lifetimes based on the data in Section 8.5.1, then based on our sample we calculate

```
MTB > SET C1
DATA> 499 26 614 231 719 2063 2723 466
DATA> 709 904 2303 27 1374 981 654
DATA> END OF DATA
MTB > NAME C1 'BULBLIFE'
MTB > NOTE EXAMPLE 8.5.1A
MTB > MEDIAN OF 'BULBLIFE'
   MEDIAN =      709.00
```

The sample median bulb life, and hence the estimated population median bulb life, is 709 hours.

Other Percentiles.   As discussed in Chapter 3, there is no command in Minitab to calculate general sample percentiles. Sample percentiles other than the median can be computed by ordering the data with the command

---
SORT data in C, put into C
---

which orders the values the column, putting the smallest value at the top of the result column, and then selecting the appropriate row of the sorted column. The total number of rows n in the result column can be obtained by

---
COUNT data in C, put result in **K**
---

which when multiplied by the desired percentage p gives (approximately) the correct row $k = pn$ to use to estimate the sample percentile. The sample percentile is then taken as an estimate of the population percentile.

For instance, if we are interested in estimating the upper population quartile (75th percentile) of bulb lifetimes based on our example data set, we are approximately correct to proceed as follows:

```
MTB > SORT 'BULBLIFE' PUT C10
MTB > COUNT 'BULBLIFE' PUT K1
   COUNT   =      15.000
MTB > NOTE AN APPROXIMATE FORMULA
MTB > LET K2 = ROUND(.75*K1)
MTB > LET K3 = C10(K2)
MTB > PRINT K3 (PERCENTILE)
K3        981.000
```

In the LET command C10(K2) gives the value in the K2th row of C10. The ROUND function, which corresponds to the command

---
ROUND values in **E**, put in **C**
---

and rounds values to the nearest integer, is used to make sure we have an integer row number. (With this function, values exactly between integers are rounded to the next larger absolute value,i.e. 12.5 goes to 13, and -3.5 goes to -4).

Unfortunately $k = pn$ is not quite the right index to use for this estimation, as it is not symmetrical in its results between the upper and lower percentiles. A better choice is $k = p(n + 1)$, as in the following:

```
MTB > NOTE A MORE SYMMETRIC FORMULA
MTB > LET K2 = ROUND(.75*(K1+1))
MTB > LET K3 = C10(K2)
MTB > PRINT K3 (PERCENTILE)
K3         1374.00
```

But neither of these quite match the definition based on the sample cumulative distribution, which requires the average of two sample values in cases where there is an observed value that makes the cumulative distribution function exactly the desired percentage. The following will do the trick:

```
MTB > NOTE THE CORRECT FORMULA
MTB > LET K2 = ROUND(.75*K1+.5)
MTB > LET K3 = K1 - ROUND((1-.75)*K1-.5)
MTB > LET K4 = (C10(K2)+C10(K3))/2
MTB > PRINT K4 (PERCENTILE)
K4         1374.00
```

The estimated upper population quartile of bulb lifetimes is 1374 hours.

Parametric Percentile Estimates. Percentile estimates based on the normal distribution can easily be calculated from the estimates resulting from the MEAN and STDEV commands and are left as an exercise.

8.5.2   Interval Estimation of a Median

A confidence interval for the median can be calculated with the command[4]

SINTERVAL [with **K**% confidence] for **C**

This confidence interval is based on the binomial distribution of the number of values greater than the median (the "signs") of the data, and corresponds to the sign test we shall discuss in Chapter 9. For instance, for the bulb life data

```
MTB > SINTERVAL WITH 95 PERCENT CONFIDENCE FOR 'BULBLIFE'

SIGN CONFIDENCE INTERVAL FOR MEDIAN

                            ACHIEVED
                 N   MEDIAN  CONFIDENCE   CONFIDENCE INTERVAL   POSITION
BULBLIFE        15   709.0   0.8815      (  499.0,    981.0)      5
                            0.9648      (  466.0,    1374)       4
```

Because of the discrete nature of the procedure, not all confidence levels can
be achieved exactly. Here Minitab computes two intervals with confidence
levels above and below the level of interest. Here we would report we were
96% confidenct that the median lies in the interval (466, 1374).

8.6 PAIRED MEASUREMENTS

If corresponding rows of two columns contain before and after measurements on the same individual or measurements on each of a pair of
matched samples, then the commands described above can be used on the
difference of two columns (obtained by the SUBTRACT or LET commands)
to compare the two treatments.

For instance, suppose we are interested in evaluating the effect of second grade on reading scores, using the data in Table 8.4 [READING.DAT].
We compute the gain in reading scores as the difference of the before and
after reading scores. We then can compute statistics and confidence intervals on the difference in reading score with any of the methods described
above. For instance

```
MTB > READ 'READING.DAT' C1-C3
      30 ROWS READ
  ROW   C1    C2    C3

   1     1    1.1   1.7
   2     2    1.5   1.7
   3     3    1.5   1.9
   4     4    2.0   2.0
    .    .    .

MTB > NAME C1 'PUPIL' C2 'BEFORE' C3 'AFTER'
MTB > NOTE EXAMPLE 8.6.1
MTB > NOTE COMPUTE MEAN READING SCORE DIFFERENCE
MTB > LET C4 = 'AFTER' - 'BEFORE'
MTB > NAME C4 'DIFF'
MTB > MEAN 'DIFF'
      MEAN    =      0.51000
MTB > TINTERVAL WITH 95 PERCENT CONFIDENCE FOR 'DIFF'

                N    MEAN    STDEV   SE MEAN    95.0 PERCENT C.I.
DIFF           30   0.5100   0.4915   0.0897   ( 0.3264,   0.6936)
```

Therefore the estimated mean reading score increase among second grade pupils is 0.49, and a 95% confidence interval for the increase is from 0.33 to 0.69.

COMMAND REFERENCE

Here are the complete forms of commands used in this chapter:

MEAN of the values in C, [put into **K**]

STDEV of C, [put standard deviation into **K**]

(Or the alternative command STANDARD.)

ZINTERVAL [with **K**% confidence] assuming $\sigma = $ **K**, for data in C,...,C

TINTERVAL [with **K**% confidence] for data in C,...,C

PDF, CDF, INVCDF

T with **K** degrees of freedom

CODE (**K**,...,**K**) to **K** (**K**,...,**K**) to **K** ... in C,...,C, put into C,...,C

MEDIAN of the values in C [put into **K**]

SORT data in C, [carry along C,...,C] , put into C,...,C

COUNT data in C, put result in **K**

ROUND values in E, put in E

SINTERVAL [with **K**% confidence] for C

COMPUTER EXERCISES

8.1 Assume that the incomes of the 49 families given in Table 2.16 [IN-COME.DAT] represent a random sample of incomes in a given city.

(a) Read the data set into Minitab. Convert to dollars by multiplying by 100.

(b) Calculate estimates of the mean, median, standard deviation, and variance of incomes in the city based on this sample.

(c) Calculate 95% confidence intervals for the population mean and median income.

(d) Estimate the population 90th percentile of incomes in the city, based on the cumulative distribution. Now construct an estimate of the 90th percentile assuming the data are normally distributed. Compare the two results. Plot a histogram of the values, and comment on the applicability of the normal assumption.

8.2 We can use Minitab not only to compute statistical results from data, but also to test the effectiveness of statistical procedures in known situations, using simulated data.

(a) Using the RANDOM command we learned in Chapter 7, generate a set of 30 values from a normal distribution with a mean of 10 and a standard deviation of 5. Compute confidence intervals for the population mean based on the 30 values, using both the z-interval (assuming we know the standard deviation is 5) and the t-interval method. Do this 20 times, and see how many times the confidence intervals actually cover the true mean of 10. (You may want to use the STORE command to set up a loop).

(b) Suppose we used the wrong value of σ in constructing our z-intervals above? Repeat the trials, and see what happens to the coverage probability of the z-interval if we assume the standard deviation is 4, when the true value is 5.

8.3 Using the light bulb lifetime data in [BULB.DAT], compute 90% confidence intervals for the mean and median of bulb lifetime, using the procedures discussed in this Chapter.

8.4 Using the before and after IQ data on heart patients in [HEART.DAT], compute 95% confidence intervals for the mean and median effect of surgery on IQ change during the disease. What is the minimum effect on the change in IQ we can attribute to the effect of surgery with 95% confidence? Does surgery appear to be recommended in such cases?

NOTES

1. Because this is a two-sided interval, to find z from the table in Appendix II we look up a value of $\Phi(z) = 1 - \frac{1-K/100}{2}$.

2. Because this is a two-sided interval, the value of t_{n-1} can be obtained from the table in Appendix III by taking $\alpha = 1 - \frac{K/100}{2}$ and $f = n-1$.

3. In previous versions of Minitab the appropriate command is

RECODE values from **K** [to **K**] in **C** to **K** put into **C**

4. The SINTERVAL command is not available in earlier versions of Minitab. The techniques used in Section 8.5.1 for estimation of percentiles

can be used to calculate interval estimates for the median, using the approximate formula given in the text. We calculate

$$r = \frac{n+1}{2} - z\frac{\sqrt{n}}{2},$$

where z is the appropriate normal value for a two-sided interval at the desired level, and round to the next lower integer. The interval $(x_{(r)}, x_{(n-r+1)})$, obtained from the ordered sample, is a distribution-free confidence interval for the population median.

9 ANSWERING QUESTIONS ABOUT POPULATION CHARACTERISTICS

INTRODUCTION

One of the most important uses of a statistical computer package is in testing statistical hypotheses. This chapter describes the available commands in Minitab for testing hypotheses about the mean of a population.

9.1 DECIDING WHETHER A POPULATION MEAN IS LARGE

The Minitab command to test hypotheses about the mean of data from a normal distribution with known standard error is[1]

ZTEST of μ =K, assuming σ =K on C

with subcommand

ZTEST
ALTERNATIVE = K

The alternative parameter determines the particular hypothesis to be tested:

Alternative = 1	H:	$\mu \leq \mu_0$	against	Alternative:	$\mu > \mu_0$
Alternative = 0	H:	$\mu = \mu_0$	against	Alternative:	$\mu \neq \mu_0$
Alternative = -1	H:	$\mu \geq \mu_0$	against	Alternative:	$\mu < \mu_0$

Using the IQ data on second-grade pupils in Table 2.23 [IQ.DAT] let us ask whether the mean math IQ of second graders is greater than 95. (The null hypothesis is that the mean is less than or equal to 95, and the alternative is that the mean is greater than 95.) Suppose further that we know the population standard deviation of math IQ is 10 points and has a normal distribution. Then we test the hypothesis as follows:

```
MTB > ZTEST OF MU=95 ASSUMING SIGMA=10 ON 'MATHIQ';
SUBC> ALTERNATIVE = 1.

TEST OF MU =  95.00 VS MU G.T.  95.00
THE ASSUMED SIGMA = 10.0

                N       MEAN    STDEV   SE MEAN      Z    P VALUE
MATHIQ         23     100.30    11.60      2.09    2.54    0.0056
```

The number of pupils in the sample is printed, with the mean and standardard deviation of math IQ. The standardized score

$$z = \frac{\overline{x} - \mu_0}{\sigma/\sqrt{n}},$$

is printed, along with the achieved significance level corresponding to z from a table of the standard normal. The level of achieved significance is the probability area in the tail of the distribution corresponding to the particular alternative. (In this case the probability of a standard normal variate being greater than 2.54.) If the level of achieved significance is smaller than the desired level of the test (in principle, chosen before the test results are examined), then the test is significant at that level. For instance, if the desired test level is 0.05, (a common level of significance), then since the achieved level of 0.0056 is less than 0.05, the mean math IQ of second- graders is found to be significantly greater than 95.

If we had posed the opposite hypothesis, that the mean math IQ is less than 95, we would have obtained the following:

```
MTB > ZTEST OF MU=95 ASSUMING SIGMA=10 ON 'MATHIQ';
SUBC> ALTERNATIVE = -1.

TEST OF MU =  95.00 VS MU L.T.  95.00
THE ASSUMED SIGMA = 10.0

                N       MEAN    STDEV   SE MEAN      Z    P VALUE
MATHIQ         23     100.30    11.60      2.09    2.54    0.99
```

Here the null hypothesis is that the mean IQ is greater than or equal to 95, versus the alternative that the mean IQ is less than 95. The z score is the same, but the achieved level of significance is the probability that a normal variate is less than 2.54. This number is greater than 0.5 and cannot be less than any reasonable desired significance level. Based on this analysis, we conclude that it is not possible to reject the null hypothesis that the mean IQ is greater than or equal to 95.

9.2 DECIDING WHETHER A POPULATION MEAN DIFFERS FROM A GIVEN VALUE

To test the null hypothesis that the mean math IQ is equal to 95 versus the alternative that it is different from 95, we do as follows:

```
MTB > ZTEST OF MU=95 ASSUMING SIGMA=10 ON 'MATHIQ';
SUBC> ALTERNATIVE = 0.

TEST OF MU =  95.00 VS MU N.E.  95.00
THE ASSUMED SIGMA = 10.0

              N      MEAN    STDEV   SE MEAN      Z    P VALUE
MATHIQ       23    100.30    11.60     2.09     2.54    0.011
```

The achieved significance for this two-tailed test is the probability that a normal random variable is greater than $|z|$ or is less than $-|z|$. (Notice this is simply twice the probability of exceeding $|z|$.) If our desired level of significance is 0.05, we reject the null hypothesis that the mean math IQ is 95. If our desired level is 0.01, we do not reject the null hypothesis.

The ALTERNATIVE subcommand is optional, and a two-sided test is assumed if no alternative is specified. For instance, to test the hypothesis that the mean math IQ is 100, we do

```
MTB > ZTEST OF MU=100, ASSUMING SIGMA=10 ON 'MATHIQ'

TEST OF MU = 100.00 VS MU N.E. 100.00
THE ASSUMED SIGMA = 10.0

              N      MEAN    STDEV   SE MEAN      Z    P VALUE
MATHIQ       23    100.30    11.60     2.09     0.15     0.88
```

The achieved siginificance is large, and we cannot reject the hypothesis that the mean math IQ is 100.

If the hypothesized mean is not specified, a test of mean = 0 is performed.

9.3 TESTING HYPOTHESES ABOUT A MEAN WHEN THE STANDARD DEVIATION IS UNKNOWN

If the population standard deviation is unknown, hypothesis tests about the mean of a normal population are performed with the command[2]

> **TTEST** [of $\mu = $ **K**] on **C**

with subcommand

> *TTEST*
> **ALTERNATIVE = K**

To test the hypothesis that the mean math IQ is greater than 95 without the knowledge of the standard deviation of IQ, we do

```
MTB > TTEST OF MU=95 ON 'MATHIQ';
SUBC> ALTERNATIVE = 1.

TEST OF MU =  95.00 VS MU G.T.   95.00

                N      MEAN    STDEV   SE MEAN       T    P VALUE
MATHIQ         23    100.30    11.60      2.42    2.19      0.020
```

The t score

$$t = \frac{\bar{x} - \mu_0}{s/\sqrt{n}}$$

is printed along with the achieved level of significance from a t-table with $n - 1$ degrees of freedom, based on the specified alternative. Based on this test, we reject the null hypothesis that the population mean math IQ is less than or equal to 95 at level 0.05.

As with ZTEST, if no alternative is specified, a two-sided alternative is assumed. For instance

```
MTB > TTEST OF MU=95 ON 'MATHIQ'

TEST OF MU =  95.00 VS MU N.E.   95.00

                N      MEAN    STDEV   SE MEAN       T    P VALUE
MATHIQ         23    100.30    11.60      2.42    2.19      0.039
```

If the hypothesized mean is not specified, a test of mean = 0 is performed.

9.4 TESTING HYPOTHESES ABOUT A PROPORTION

The general computational abilities of Minitab allow the testing of hypotheses about proportions according to the methods described in the text. For instance, working the occupational data from Table 2.1, we may be interested in testing the null hypothesis that the population proportion of professionals is less than 0.6. As in Section 8.4, we create a column with a 1 for professionals and a 0 for others. Then we compute the standardized variable

$$\frac{\hat{p} - p_0}{\sqrt{p_0 q_0/n}},$$

and refer to a table of the standard normal distribution to determine the significance.

```
MTB > NOTE NOW CALCULATE TEST STATISTIC
MTB > NOTE FOR THE POPULATION PROPORTION OF PROFESSIONALS
MTB > MEAN 'PROFESS' , RESULT IN K1
   MEAN     =      0.30000
MTB > COUNT 'PROFESS', RESULT IN K2
   COUNT    =      20.000
MTB > LET K3 = (K1 - .6)/SQRT(.6*.4/K2)
MTB > PRINT K3
K3        -2.73861
```

Using the table in Appendix II, the standardized variable is less than -1.645, which corresponds to a one-sided normal test at level 0.05. We reject the hypothesis that the population proportion of professionals is 0.6 in favor of the alternative that it is less than 0.6.

A more precise procedure for calculating the standardized variable is to apply the continuity correction as described in Appendix 9A, and compute the continuity-corrected standardized variable

$$\frac{\hat{p} - p_0 - \text{sign}(\hat{p} - p_0)\frac{1}{2n}}{\sqrt{p_0 q_0 / n}},$$

as follows:

```
MTB > NOTE WITH CONTINUITY CORRECTION
MTB > LET K4 = (K1 - .6 - SIGN(K1-.6)/(2*K2))/SQRT(.6*.4/K2)
MTB > PRINT K4
K4        -2.51040
```

Based on this calculation, we reject the null hypothesis at level 0.05, as the continuity-corrected standardized score is less than -1.645.

9.5 TESTING HYPOTHESES ABOUT A MEDIAN: THE SIGN TEST

Hypotheses about the median can be tested with the command[3]

STEST sign test [median = **K**] for data in **C**

with subcommand

STEST
ALTERNATIVE = K

For instance, returning to the math IQ data, we can test whether the median IQ is greater than 95 by the sign test, as follows

```
MTB > STEST OF MEDIAN=95 ON 'MATHIQ';
SUBC> ALTERNATIVE = 1.

SIGN TEST OF MEDIAN = 95.00 VERSUS  G.T.  95.00

                 N  BELOW  EQUAL  ABOVE   P-VALUE    MEDIAN
MATHIQ          23     10      0     13    0.3388     97.00
```

The median is not found to be significantly different from 95.

9.6   PAIRED MEASUREMENTS

9.6.1   Testing Hypotheses About the Mean of a Population of Differences

If we have paired measurements and are interested in the mean difference of the measurements, we compute the difference for each pair in the sample, and apply the procedures described above. For instance, with the before and after reading scores from Table 8.4 we have

```
MTB > LET C4 = 'AFTER' - 'BEFORE'
MTB > NAME C4 'DIFF'
MTB > TTEST FOR MU=0 ON 'DIFF';
SUBC> ALTERNATIVE=1.

TEST OF MU = 0.0000 VS MU G.T. 0.0000

             N     MEAN   STDEV  SE MEAN        T   P VALUE
DIFF        30   0.5100  0.4915   0.0897     5.68    0.0000
```

A highly significant increase in reading scores is found.

9.6.2   Testing the Hypothesis of Equality of Proportions

As with simple proportions, testing the equality of proportions with Minitab is a matter of using the formulas given in the text. If the data are available in a case-by-case form, they can be summarized with the TABLE command, as discussed in Chapter 5. The off-diagonal frequencies (B and C) can be taken from this table by hand, and a standardised test statistic can be computed for testing that the population proportion corresponding to $B/(B + C)$ is equal to .5.

9.6.3   The Sign Test for the Median of Differences

The data in Table 9.20 on civil liberties scores for campers at the beginning and end of the summer can be analyzed using a data set [CAMPER.DAT] with one entry for each camper containing the before and after score (i.e. 10 cases with 0 and 0, 2 cases with 0 and 1, etc.). Using these data we can perform the calculations to test whether the median difference in the civil liberties score after a summer encampment is nonzero.

```
MTB > NOTE INDIVIDUAL CHANGES IN CAMPER'S SCORES ON CIVIL LIBERTIES
MTB > NOTE DURING SUMMER
MTB > READ 'CAMPER.DAT' C1-C2
      96 ROWS READ
  ROW    C1    C2

    1     0     0
    2     0     0
    3     0     0
    4     0     0
    .     .     .

MTB > NAME C1 'BEGIN' C2 'END'
MTB > LET C3 =  'END' - 'BEGIN'
MTB > NAME C3 'CHANGE'
MTB > STEST OF MEDIAN DIFFERENCE=0 FOR 'CHANGE'

SIGN TEST OF MEDIAN = 0.000000000 VERSUS  N.E.  0.000000000
```

	N	BELOW	EQUAL	ABOVE	P-VALUE	MEDIAN
CHANGE	96	48	35	13	0.0000	-0.5000

A highly significant decrease in the score (improvement on the test) is found.

APPENDIX 9B THE SIGNED RANK TEST

The Wilcoxon signed rank test and associated confidence procedure are implemented with the commands

WTEST [of center = **K**] on **C**

with subcommands

WTEST
ALTERNATIVE = **K**

and

WINTERVAL [with **K**% confidence] for data in **C**

These perform the signed rank procedure as described in the text. The confidence interval at level **K** is the set of values for the center that are not rejected by the signed rank test at the 1 - **K**/100 level. If the center is not specified on the WTEST command, the test is of whether the center is 0. Of course, both of these commands can be applied to evaluating the differences of paired measurements in the same way as TTEST and TINTERVAL were above.

```
MTB > WTEST OF CENTER=95 ON 'MATHIQ'

TEST OF MEDIAN = 95.00 VERSUS MEDIAN N.E. 95.00

                 N FOR   WILCOXON              ESTIMATED
            N    TEST   STATISTIC  P-VALUE      MEDIAN
MATHIQ     23     23      193.0    0.097        99.00
MTB > WINTERVAL ON 'MATHIQ'

                ESTIMATED   ACHIEVED
            N    MEDIAN    CONFIDENCE   CONFIDENCE INTERVAL
MATHIQ     23    99.00       95.0     (  94.50,  105.50)
```

COMMAND REFERENCE

Listed below are the general forms of the Minitab commands discussed in this chapter.

ZTEST of $\mu = K$, assuming $\sigma = K$ on C,...,C
    ALTERNATIVE = **K**
TTEST [of $\mu = K$] on C,...,C
    ALTERNATIVE = **K**
STEST sign test [median = **K**] for data in C,...,C
    ALTERNATIVE = **K**
WTEST [of center = **K**] on C,...,C
    ALTERNATIVE = **K**
WINTERVAL [with **K**% confidence] for data in C

COMPUTER EXERCISES

9.1 Read in the tensile strength data from Table 9.22 [TENSILE.DAT]. Assuming normality, test at the 10% level the hypothesis that the mean tensile strength of the lot is at most 60,000 psi against the alternative that it exceeds 60,000 psi.

9.2 Read in the weights of the 25 packages of hamburger meat from Table 9.23 [HAMBURGER.DAT]. Compute a two-tailed, 5%-level test of the hypothesis that the population mean weight of hamburger packages is equal to 3 pounds against the alternative that it is not equal to 3 pounds.

9.3 As noted in Exercise 9.3 of the text, it is reasonable to assume approximate normality for the sample mean difference in scores in the Hyman, Wright, and Hopkins study (Table 9.20) because of the large sample size. Read in the data from this study in extensive form [CAMPER.DAT]. Test at the 5% level the hypothesis that the population mean difference in score is zero againset the one-sided alternative that the population mean difference is negative (indicating improvement).

9.4 Read in the data in [BEANS.DAT] on the number of beans sorted by two methods by nine workers. Test whether method two is better than method one, using

- (a) *t*-test.
- (b) sign test.
- (c) signed rank test.

Comment on the results.

NOTES

1. In older versions of Minitab, the alternative is specified on the command line, as follows:

ZTEST of $\mu = $**K**, [alternative = **K**] assuming $\sigma = $**K** on **C**

2. In older versions

TTEST of $\mu = $ **K** [alternative = **K**,] on **C**

3. The STEST command is not available in earlier Minitab versions. As with hypotheses about proportions, we can use the general calculation abilities of Minitab to calculate the sign test for the median. We use the SIGNS function to obtain +1, 0 or -1 according to whether the difference between the variable and and the hypothesized value is greater than, equal to, or less than 0. Then we rescale the results to get a 0/1 variable indicating whether the difference is greater than or less than the hypothesized value. Under the hypothesis, values will be greater than or less than the hypothesized median with equal probability, and we can perform a test that the proportion of such cases is equal to .5 to test the hypothesis about the median.

10 DIFFERENCES BETWEEN TWO POPULATIONS

INTRODUCTION

Minitab has several commands for comparing the means of two groups. When measurements are entered into Minitab for two or more groups, there are two approaches: the data for each group can be entered in a separate column, or the measurements can be all in one column, with another categorical variable indicating the group membership. In the latter case, plots of the data for each group can be obtained from the HISTOGRAM and DOTPLOT command, using the subcommand

```
HISTOGRAM, DOTPLOT
   BY C
```

which requests separate plots for each level of the by-column. If plots with comparable scales are desired, the subcommand

```
HISTOGRAM, DOTPLOT
   SAME scales
```

can be used.

10.1 COMPARISON OF TWO INDEPENDENT SAMPLE MEANS WHEN THE POPULATION STANDARD DEVIATIONS ARE KNOWN

10.1.1 One-Tailed Tests

When the population standard deviations are known, we compute the standardized difference between the two groups based on the differences of the group means and the sample sizes. For instance, to test whether the means of group A and group B on the standardized test are the same, based on the data from Table 10.1, we set up the data for each group in a column, as follows:

```
MTB > SET C1
DATA> 97 83 81
DATA> 95 85 90
DATA> 86 110 121
DATA> 102 119 105
DATA> 119 117 104
DATA> 99 108 96
DATA> 87 74
DATA> 101 93
DATA> END OF DATA
MTB > SET C2
DATA>  89 104 107 85 70 91 96
DATA> END OF DATA
MTB > NAME C1 'GP-A' C2 'GP-B'
MTB > DESCRIBE C1 C2
```

	N	MEAN	MEDIAN	TRMEAN	STDEV	SEMEAN
GP-A	22	98.73	98.00	98.85	13.35	2.85
GP-B	7	91.71	91.00	91.71	12.43	4.70

	MIN	MAX	Q1	Q3
GP-A	74.00	121.00	86.75	108.50
GP-B	70.00	107.00	85.00	104.00

Then we compute the difference of means, and the standard deviation of the difference of the means, assuming that the population standard deviation in each subgroup is 12.

```
MTB > LET K1 = MEAN('GP-A') - MEAN('GP-B')
MTB > LET K2 = 12 * SQRT(1/COUNT(C1) + 1/COUNT(C2))
MTB > LET K3 = K1/K2
MTB > PRINT K3
K3      1.34674
```

The standardized difference of 1.347 is greater than the one-sided percentage point of the normal distribution corresponding to a 10% one sided test (1.282), so the null hypothesis is rejected.

10.1.2 Two-Tailed Tests

The same procedure is followed for two-sided tests, except the standardized difference is compared to normal values corresponding to a two-sided test.

10.1.3 Confidence Intervals

Confidence intervals for the differences of means are calculated using the appropriate normal value and the standard deviation of the difference. For instance, continuing the example above, a 95% confidence interval for the difference of the means is calculated as follows.

```
MTB > LET K4 = K1 - 1.96*K2
MTB > LET K5 = K1 + 1.96*K2
MTB > PRINT K4 K5
K4        -3.19349
K5        17.2195
```

We are 95% confident that the difference between the means of group A and group B lies in the interval (-3.19, 17.22).

10.2 COMPARISON OF TWO INDEPENDENT SAMPLE MEANS WHEN THE POPULATION STANDARD DEVIATIONS ARE UNKNOWN BUT TREATED AS EQUAL

When the population standard deviations are unknown, but are assumed to be equal in each group, a pooled estimate of the standard deviation is used to compute the standardized difference. This is compared with a Student's t-distribution in order to test the significance of the difference. The Minitab command to perform this test of the difference of means is[1]

TWOSAMPLE t-test on C vs C
---

with subcommands

TWOSAMPLE

ALTERNATIVE = K
---

and

TWOSAMPLE

POOLED

The data for each group is stored in its own column. The alternative is specified in the same way as for the TTEST and ZTEST commands described in Chapter 8 – a value of +1 specifies a one-sided test of $\mu_1 \leq \mu_2$ versus $\mu_1 > \mu_2$, a value of -1 specifies a test of $\mu_1 \geq \mu_2$ versus $\mu_1 < \mu_2$, and a value of 0 specifies a test of $\mu_1 = \mu_2$ versus $\mu_1 \neq \mu_2$. The POOLED subcommand specifies the two standard deviations are to be treated as equal. For example

```
MTB > TWOSAMPLE T TEST FOR 'GP-A' VS 'GP-B';
SUBC> ALTERNATIVE=0;
SUBC> POOLED.

TWOSAMPLE T FOR GP-A VS GP-B
```

```
          N     MEAN    STDEV   SE MEAN
GP-A   22     98.7    13.3      2.8
GP-B    7     91.7    12.4      4.7

95 PCT CI FOR MU GP-A - MU GP-B: (-4.7, 18.7)
TTEST MU GP-A = MU GP-B (VS NE): T=1.23 P=0.23 DF=27.0
```

The standardized difference is 1.23, which corresponds to a two-tailed probability of 0.23 according to a t-distribution with 27 degrees of freedom. Therefore the test is not significant at the 0.05 level.

If the data are all stored in one column, with a separate column containing group codes, the command

---
TWOT *t*-test on **C**, codes in **C**
---

produces an equivalent analysis and takes the same subcommands. For instance, to test whether the mean score of undergraduates is equal to the mean score of graduates based on the final grades in Table 10.4, we read in the file [STATGRAD.DAT] containing the final grades. The first value on each line is the student number, the second value is the class (1 = undergraduate, 2 = graduate), and the third value is the final score.

```
MTB > READ 'STATGRAD.DAT' C1-C3
      68 ROWS READ
 ROW    C1    C2    C3

   1     1     1   153
   2     2     1   109
   3     3     1   157
   4     4     1   145
     .   .   .

MTB > NAME C1 'STUDENT' C2 'CLASS' C3 'SCORE'
MTB > NOTE CLASS: UNDERGRADUATE=1, GRADUATE=2
```

Then to test whether the means are equal, we do

```
MTB > TWOT FOR DATA IN 'SCORE', CODES IN 'CLASS';
SUBC> ALTERNATIVE=0;
SUBC> POOLED.

TWOSAMPLE T FOR SCORE
CLASS    N     MEAN    STDEV   SE MEAN
1       39    136.4    20.2      3.2
2       29    127.0    21.9      4.1

95 PCT CI FOR MU 1 - MU 2: (-0.8, 19.7)
TTEST MU 1 = MU 2 (VS NE): T=1.84 P=0.071 DF=66.0
```

The standardized difference is 1.84, which corresponds to a two-tailed probability of 0.071 according to a *t*-distribution with 66 degrees of freedom. Therefore the test is not significant at the 0.05 level.

As with ZTEST and TTEST, the default alternative value is 0, if not specified. Descriptive statistics for each group are printed, as is a confidence interval for the difference of the means. The confidence level can be specified as part of the commands as follows:

TWOSAMPLE *t*-test and c.i. [with **K%** confidence] on **C** vs **C**

TWOT *t*-test and c.i. [with **K%** confidence] on **C**, codes in **C**

For instance, if we want a 90% confidence interval for the difference in scores, we do

```
MTB > TWOT WITH 90 PERCENT CI, FOR DATA IN 'SCORE', CODES IN 'CLASS';
SUBC> POOLED.

TWOSAMPLE T FOR SCORE
CLASS    N       MEAN     STDEV    SE MEAN
1       39       136.4    20.2       3.2
2       29       127.0    21.9       4.1

90 PCT CI FOR MU 1 - MU 2: (0.9, 18.0)
TTEST MU 1 = MU 2 (VS NE): T=1.84 P=0.071 DF=66.0
```

The default confidence level is 95%, if not specified. Note that in the previous example, the 95% confidence interval did include 0, correspondingly the test would not reject $\mu_1 - \mu_2 = 0$ at level 0.05. In this example, the 90% confidence interval does not include 0, which corresponds to rejecting the hypothesis of equality of the means at level 0.10.

10.3 COMPARISON OF TWO INDEPENDENT SAMPLE MEANS WHEN THE POPULATION STANDARD DEVIATIONS ARE UNKNOWN AND NOT TREATED AS EQUAL

In cases where the population standard deviations are unknown and may be unequal the TWOSAMPLE and TWOT commands (without the POOLED subcommand) calculate the standardized difference and confidence interval using the unpooled estimate of the standard deviation of the difference

$$\sqrt{\frac{s_1^2}{n_1} + \frac{s_2^2}{n_2}}.$$

The standardized difference of the means has an approximate *t*-distribution, with degrees of freedom depending on both the sample sizes in the groups and the relative size of the standard deviations of the two groups. This approximation, known as Welch's approximation, is appropriate even for considerably different standard deviations and sample sizes. It also gives almost exactly the same results as the pooled test when the group standard deviations are equal; most of the time it will be preferred to the pooled statistic, if little is known about the standard deviations.

With our example data we have:

```
MTB > TWOT FOR DATA IN 'SCORE', CODES IN 'CLASS'

TWOSAMPLE T FOR SCORE
CLASS   N       MEAN    STDEV   SE MEAN
1       39      136.4   20.2    3.2
2       29      127.0   21.9    4.1

95 PCT CI FOR MU 1 - MU 2: (-1.0, 19.8)
TTEST MU 1 = MU 2 (VS NE): T=1.82 P=0.075 DF=57.7
```

10.4 COMPARISON OF TWO INDEPENDENT SAMPLE PROPORTIONS

Comparisons of two sample proportions are easily performed using Minitab's calculating commands, according to the formulas in the text. See, also, Section 9.4 and Appendix 10A.

10.5 TEST FOR LOCATION BASED ON RANKS OF THE OBSERVATIONS

10.5.1 Sign Test for Comparing Locations

The sign test can be used for comparing data from two populations, as described in the text. The command[2]

> **STACK C on C, put into C**

is used to combined the data from the two columns so that the overall median can be computed. Then the proportion of data in each group above and below the median are compared. For instance, with the data from Table 10.8 on incomes of Protestants and Catholics, we have

```
MTB > NOTE INCOMES (100'S OF $) OF 13 PROTESTANTS AND 10 CATHOLICS
MTB > SET C1
DATA> 41,53,64,67,84,100,115,127,131,145,200,280,500
DATA> END
MTB > NAME C1 'PROTES'
MTB > SET C2
DATA> 43,47,54,61,69,78,94,121,174,260
DATA> END
MTB > NAME C2 'CATHOL'
MTB > STACK 'PROTES' ON 'CATHOL', PUT IN C3
MTB > MEDIAN OF C3, PUT IN K1
   MEDIAN =        94.000
MTB > LET C4 = SIGNS('PROTES' - K1)
MTB > LET C5 = SIGNS('CATHOL' - K1)
MTB > TABLES C4


 ROWS: C4


        COUNT


  -1         5
   1         8
 ALL        13

MTB > TABLES C5

 ROWS: C5


        COUNT


  -1         6
   0         1
   1         3
 ALL        10
```

Based on these frequencies, the calculation of the test statistic proceeds exactly as in the text.

10.5.2   Rank Sum Test for Location of Two Populations

The sign test is often referred to as *nonparametric*, because it does not require a particular distributional form. A more powerful nonparametric test for comparing two distributions is the rank sum test, also known as the Wilcoxon two-sample test or the Mann-Whitney test. The command to perform this analysis is

MANN-WHITNEY test [alternative = **K**] [with **K**% c.i.] for **C** vs **C**

For instance, using the example above, we have

```
MTB > MANN-WHITNEY TEST FOR 'PROTES' VS 'CATHOL'

Mann-Whitney Confidence Interval and Test

    PROTES     N = 13    MEDIAN =      115.00
    CATHOL     N = 10    MEDIAN =       73.50
    POINT ESTIMATE FOR ETA1-ETA2 IS     23.5045
    95.6  PCT C.I. FOR ETA1-ETA2 IS (    -25,       84)
    W =    174.0
    TEST OF ETA1 = ETA2  VS.  ETA1 N.E. ETA2 IS SIGNIFICANT AT   0.2778

CANNOT REJECT AT ALPHA = 0.05
```

ETA1 and ETA2 are the medians of the two distributions; the difference between ETA1 and ETA2 is estimated and a confidence interval is calculated.
Alternatives are the same as for the other commands in this chapter, and
the two-sided alternative (ALTERNATIVE = 0) is assumed if not specified.
A 95% confidence level is assumed if none is specified, as above. Because
the confidence bounds must occur at differences of data values, not all confidence levels can be achieved exactly. The closest available level above 95%
in this example is 95.6%, so a 95.6% confidence interval is reported.

COMMAND REFERENCE

Listed below are the general forms of the Minitab commands discussed in this chapter.

HISTOGRAM for data in C,...,C
    BY C
    SAME scales
DOTPLOT for data in C,...,C
    BY C
    SAME scales
TWOSAMPLE *t*-test and c.i. [with **K**% confidence] on **C** vs **C**
    ALTERNATIVE = K
    POOLED
TWOT *t*-test and c.i. [with **K**% confidence] on **C**, codes in **C**
    ALTERNATIVE = K
    POOLED
STACK C on **C**, put into **C**
MANN-WHITNEY test [alternative = **K**] [with **K**% c.i.] for **C** vs **C**

COMPUTER EXERCISES

10.1 Using the yield data on apple trees and the effect of fungicide in [APPLE.DAT]:

(a) Obtain histograms of the yield for sprayed and unsprayed trees.

(b) Perform a *t*-test of whether the yield of the sprayed trees is greater than the unsprayed trees, assuming the variances of the two populations are equal.

(c) Perform a *t*-test of the same item without assuming the variances are equal.

(d) Test the same hypothesis with the sign test.

(e) Test the same hypothesis with the rank-sum test.

10.2 Using the data on the two carbonate environments in [CARBON.-DAT], obtain dotplots of the four measurement variables for each environment using the same scale for each pair. Comment on the apparent differences in the two environments. Perform a rank-sum test on each measurement to test for differences.

10.3 Using the data on ingredients of cookie and cake recipes in [COOKIE.DAT], obtain comparable plots of the data for each ingredient for each group. Perform nonparametric tests of the differences in each ingredient between cookies and cakes. Summarize your results.

NOTES

1. In earlier versions of Minitab, the two-sample *t*-test is performed with

POOLED *t*-test [alternative = **K**] [with **K%** c. i.] on **C** vs **C**

for a test with standard deviations treated as equal, and

TWOSAMPLE *t*-test [alternative = **K**] [with **K%** c. i.] on **C** vs **C**

for a test with standard deviations not treated as equal.

2. In earlier versions of Minitab, the command to use is

JOIN C on C, put into C

11 VARIABILITY IN ONE POPULATION AND IN TWO POPULATIONS

INTRODUCTION

Minitab does not have any specific commands for testing equality of variances or calculating confidence intervals for variances. However the procedures described in this chapter can all be performed using the functions and commands we have already learned with the aid of tables of the chi-square and F distributions.

11.1 VARIABILITY IN ONE POPULATION

11.1.2 Testing the Hypothesis that the Variance Equals a Given Number

In Chapter 9 we assumed that the population standard deviation of math IQ's among second graders was 10. To test this hypothesis we can calculate the test statistic $(n-1)s^2/\sigma^2$ and compare it with valuses in a table of the chi-square distribution with $n-1$ degrees of freedom, such as Appendix IX.

```
MTB > LET K1 = STDEV('MATHIQ')**2
MTB > LET K2 = COUNT('MATHIQ')-1
MTB > LET K3 = K2*K1/100
MTB > PRINT K1,K2,K3
K1      134.585
K2       22.0000
K3       29.6087
```

The test value of 29.6 is neither larger than $\chi^2_{22}(.025) = 36.781$ nor smaller than $\chi^2_{22}(.975) = 10.982$, so the hypothesis that the population standard deviation is 10 (variance of 100) is not rejected at the 5% level.

The subcommand

RANDOM, PDF, CDF, INVCDF
   CHISQUARE with **K** degrees of freedom
---

can be used with the PDF, CDF, and INVCDF to obtain chi-square distribution values. For instance, in the above example, we can compute

```
MTB > CDF FOR K3;
SUBC> CHISQUARE WITH K2 DF.
    29.61    0.8717
```

Since the cumulative distribution is neither less than 0.025 nor greater than 0.975, the test is not significant at level 0.05.

11.1.3   Confidence Intervals for the Variance

A 95% confidence interval for the variance can be calculated using the same values from the chi-square table, as follows:

```
MTB > LET K4 = K2*K1/36.781
MTB > LET K5 = K2*K1/10.982
MTB > PRINT K4,K5
K4        80.5000
K5        269.611
```

We are 95% confident that the population variance lies between 80.5 and 269.6. Taking square roots, we can say that we are 95% confident that the population standard deviation lies between 8.97 and 16.42.

11.2   VARIABILITY IN TWO POPULATIONS

11.2.2   Testing the Hypothesis of Equality of Two Variances

The ratio of two variances can be computed and compared to values in a table of the F-distribution, such as Appendix X. For instance, with the final grades from Table 10.4 [STATGRAD.DAT] we may wish to test whether the population variances of the scores for the two groups are equal. Since the values are all in 'SCORE', with the class indicated by 'CLASS', we first need to place the scores into separate columns for each class, as follows

```
MTB > UNSTACK 'SCORE' INTO C4,C5;
SUBC> SUBSCRIPTS IN 'CLASS'.
MTB > NAME C4 'UGRAD' C5 'GRAD'
```

The command

UNSTACK C into C,...,C

> *UNSTACK*
>> **SUBSCRIPTS** are in C

separates the 'SCORE' values according to the values of 'CLASS'.
    Then we compute the F-ratio for the variances

```
MTB > LET K1 = STDEV('UGRAD')**2/STDEV('GRAD')**2
MTB > LET K2 = COUNT('UGRAD')-1
MTB > LET K3 = COUNT('GRAD')-1
MTB > PRINT K1,K2,K3
K1        0.855921
K2        38.0000
K3        28.0000
```

Referring to Appendix X for $m = 38$, $n = 28$, and interpolating as described we see that this ratio is neither less than $F_{m,n}(.025)$ nor greater than $F_{m,n}(.975)$ and so the hypothesis that the variances are equal is not rejected at the 5% level.
    The subcommand

> *RANDOM, PDF, CDF, INVCDF*
>> **F** distribution with **K** and **K** degrees of freedom

to the PDF, CDF, and INVCDF commands also can be used to obtain information on the F-distribution. For instance, in the above example

```
MTB > CDF FOR K1;
SUBC> F DISTRIBUTION WITH K2 AND K3 DF.
      0.86     0.3236
```

The cumulative distribution is neither greater than 0.975 nor less than 0.025, so the hypothesis is not rejected.

COMMAND REFERENCE

Listed below are the general forms of the Minitab commands discussed in this chapter.

PDF, CDF, INVCDF
    CHISQUARE with **K** degrees of freedom
    F distribution with **K** and **K** degrees of freedom
UNSTACK C into C,...,C
    SUBSCRIPTS are in C

COMPUTER EXERCISES

11.1 Create a data set with the standardized test scores from Table 10.1 for group A and group B. For each group, test the hypothesis that the standard deviation of scores in the corresponding population is 10.

11.2 Using the yield data on apple trees and the effect of fungicide in [APPLE.DAT]:

(a) Estimate the standard deviation and variance of yield for sprayed and usprayed trees.

(b) Test whether the variance in yield is the same for sprayed and unsprayed trees.

PART FIVE

STATISTICAL METHODS FOR OTHER PROBLEMS

12 INFERENCE ON CATEGORICAL DATA

INTRODUCTION

We saw in Chapter 5 how to create frequency tabulations from categorical data. In this chapter we shall see how to test hypotheses about the distribution of a variable and about independence of two variables.

12.1 TESTS OF GOODNESS OF FIT

12.1.1 Two Categories–Dichotomous Data

The goodness of fit test of a hypothesized distribution of a discrete variable can be calculated using general Minitab commands. For instance, in the example of the choice of doll color from the text we are given that out of 252 trials the white doll was chosen 169 times and the black doll was chosen 83 times. Under the hypothesis that each color doll is chosen with probability 1/2, the expected number of dolls chosen of each type is 126. We can calculate the X^2 test statistic as follows:

```
MTB > SET C1
DATA> 169 83
DATA> END
MTB > LET C2 = (C1 - 126)**2/126
MTB > SUM OF C2
   SUM    =       29.349
* NOTE * ALL VALUES IN COLUMN ARE IDENTICAL
```

This value is to be compared with values in a chi-square table such as Appendix IX, using 1 degree of freedom.

12.1.2 An Arbitrary Number of Categories

The same technique can be applied to an arbitrary number of categories. If the null hypothesis is equal probabilities for all categories, then the expected number in each category can be calculated as part of the procedure. For instance, in order to test the data from Table 12.1 on deaths per month before, during, and after the birth month, we put the number of deaths per month in a column, calculate the expected number of deaths per month, and then calculate the test statistic.

```
MTB > SET C1
DATA> 24 31 20 23 34 16 26 36 37 41 26 34
DATA> END
MTB > NAME C1 'DEATH'
MTB > LET K1 = SUM('DEATH')/COUNT('DEATH') # EXPECTED NUMBER
MTB > LET K2 = SUM( ('DEATH' - K1)**2/K1 )
MTB > PRINT K2
K2      22.0690
```

The X^2 value of 22.07 is to be compared with the values in a chi-square table with 11 degrees of freedom.

12.1.3   Choosing a Chi-Square Test

If the expected numbers of outcomes are not equal in all categories, then it is easiest to read in both the observed and expected values for each category. For instance, for the results of rolling one die shown in Table 12.3, we do:

```
MTB > READ C1 C2
DATA>  5  10
DATA>  38 40
DATA>  17 10
DATA> END OF DATA
      3 ROWS READ
MTB > NOTE 5  10
MTB > NOTE 38 40
MTB > NOTE 17 10
MTB > NAME C1 'OBS' C2 'EXP'
MTB > LET K1 = SUM( ('OBS' - 'EXP')**2/'EXP')
MTB > PRINT K1
K1      7.50000
```

12.2   CHI-SQUARE TESTS OF INDEPENDENCE

12.2.1   Two-by-Two Tables

Minitab has several commands for calculating the chi-square test of independence in two-way tables, depending on the format of the data. If the data have already been tabulated, then the tabled values can be typed directly into a set of columns, and the command

CHISQUARE test of independence on table in columns C,...,C

will print out the table and calculate the X^2 statistic. For instance, with the data from Table 12.4 on the favored candidate in two polls, we would have

```
MTB > READ TABLE INTO C1-C2
DATA> 523 502
DATA> 477 498
DATA> END OF DATA
        2 ROWS READ
MTB > CHISQUARE FOR TABLE IN C1-C2

Expected counts are printed below observed counts

            C1      C2     Total
    1       523     502     1025
            512.5   512.5

    2       477     498     975
            487.5   487.5

Total       1000    1000    2000

ChiSq =   0.22 +   0.22 +
          0.23 +   0.23 = 0.88
df = 1
```

Printed in each cell are the observed and expected frequencies. Below the table are the values of $(O_{ij} - E_{ij})^2/E_{ij}$ and their sum, plus the degrees of freedom of the statistic.

If the data are in a case-by-case form with categorical codes, the command

---
| **CONTINGENCY** table analysis, row class. in C, column class. in C |
---

will tablulate the data, and then print out the table and the chi-square analysis. For instance the double dichotomy data from Table 5.1 on sex versus university division was entered one row per student, with two categorical variables.

```
MTB > NOTE SEX: F=1 M=2
MTB > NOTE UNIVERSITY DIVISION: G=1 U=2
MTB > CONTINGENCY TABLE FOR 'SEX' BY 'DIVISION'

Expected counts are printed below observed counts

Rows SEX  Columns DIVISION

            1       2      Total

    1       8       4       12
            5.3     6.7

    2       3       10      13
            5.7     7.3

Total       11      14      25
```

```
ChiSq =    1.40 +   1.10 +
           1.29 +   1.02 = 4.81
df = 1
```

The output format for the CONTINGENCY command is the same as for CHISQUARE.

As we discussed in Chapter 5, the TABLE command has a CHISQUARE subcommand that calculates the chi-square statistic for the table. With our example data above we would have:

```
MTB > TABLE FOR 'SEX' BY 'DIVISION';
SUBC> CHISQUARE.

 ROWS: SEX     COLUMNS: DIVISION

              1        2     ALL

   1          8        4      12
   2          3       10      13
 ALL         11       14      25

 CHI-SQUARE =      4.812   WITH D.F. =    1

   CELL CONTENTS --
                  COUNT
```

The CHISQUARE and CONTINGENCY commands will print a warning message if the expected frequency in any cell is less than 5.[1]

When used for a table with more than two dimensions, the CHISQUARE subcommand calculates the X^2 statistic separately for each two-way subtable in the printout.

12.2.2   Two-Way Tables in General

All of the commands described above work the same way for general two-way tables. For instance, using the draft lottery data from Table 6.1 that has been typed into a file [DRAFT.DAT], we can do

```
MTB > READ 'DRAFT.DAT' C1-C4
     12 ROWS READ
  ROW    C1     C2     C3     C4

   1      1      9     12     10
   2      2      7     12     10
   3      3      5     10     16
   4      4      8      8     14
   .      .      .

MTB > NAME C1 'MONTH' C2 'LOW' C3 'MED' C4 'HI'
MTB > CHISQUARE 'LOW' 'MED' 'HI'
```

Expected counts are printed below observed counts

	LOW	MED	HI	Total
1	9	12	10	31
	10.3	10.3	10.3	
2	7	12	10	29
	9.7	9.7	9.7	
3	5	10	16	31
	10.3	10.3	10.3	
4	8	8	14	30
	10.0	10.0	10.0	
5	9	7	15	31
	10.3	10.3	10.3	
6	11	7	12	30
	10.0	10.0	10.0	
7	12	7	12	31
	10.3	10.3	10.3	
8	13	7	11	31
	10.3	10.3	10.3	
9	10	15	5	30
	10.0	10.0	10.0	
10	9	15	7	31
	10.3	10.3	10.3	
11	12	12	6	30
	10.0	10.0	10.0	
12	17	10	4	31
	10.3	10.3	10.3	
Total	122	122	122	366

```
ChiSq =   0.17 +   0.27 +   0.01 +
          0.74 +   0.56 +   0.01 +
          2.75 +   0.01 +   3.11 +
          0.40 +   0.40 +   1.60 +
          0.17 +   1.08 +   2.11 +
          0.10 +   0.90 +   0.40 +
          0.27 +   1.08 +   0.27 +
          0.69 +   1.08 +   0.04 +
          0.00 +   2.50 +   2.50 +
          0.17 +   2.11 +   1.08 +
          0.40 +   0.40 +   1.60 +
          4.30 +   0.01 +   3.88 = 37.16
df = 22
```

12.2.3   Choosing a Chi-Square Test

As we saw in Chapter 5, the FREQUENCIES subcommand of the TABLE command can be used if the data are grouped. For instance if the data from Table 12.22 on educational level and income are read in as two categorical variables plus a frequency count, we can do

```
MTB > READ C1-C3
DATA> 1 1 30
DATA> 1 2 25
DATA> 1 3  5
DATA> 2 1  3
DATA> 2 2 17
DATA> 2 3 20
DATA> END OF DATA
     6 ROWS READ
MTB > NAME C1 'EDUCAT' C2 'INCOME' C3 'FREQ'
MTB > NOTE EDUCATION: LOW=1, HIGH=2
MTB > NOTE INCOME: LOW=1, MEDIUM=2, HIGH=3
MTB > TABLE FOR 'EDUCAT' BY 'INCOME';
SUBC> FREQUENCIES 'FREQ';
SUBC> CHISQUARE.

 ROWS: EDUCAT     COLUMNS: INCOME

              1        2        3      ALL

      1      30       25        5       60
      2       3       17       20       40
    ALL      33       42       25      100

CHI-SQUARE =   29.807   WITH D.F. =    2

   CELL CONTENTS --
              COUNT
```

12.3   MEASURES OF ASSOCIATION

As described in Chapter 5, the ϕ coefficient of association for 2×2 tables can be calculated from the X^2 statistic by the relationship $\phi^2 = X^2/n$. Minitab does not have commands to calculate the other coefficients of association based on prediction and ordering discussed in this section, nor is there any easy way to calculate them with Minitab.

COMMAND REFERENCE

Listed below are the general forms of the Minitab commands discussed in this chapter.

CHISQUARE test of independence on table in columns C,...,C
CONTINGENCY table analysis, row class. in C, column class. in C
TABLE C by C
> **FREQUENCIES** in C
> **CHISQUARE** - calculate X^2 statistic

COMPUTER EXERCISES

12.1 Using the horse racing data from Table 12.30, test whether the probability of winning is the same for all post positions.

12.2 Using the data from Table 12.42, test whether auto weight is independent of accident frequency.

12.3 Create a data set containing the information in Table 12.34 on income level for two suburbs in three age groups.

> (a) Use the TABLE command to produce a table like Table 12.34, using the LAYOUT command as discussed in Chapter 5.

> (b) Test whether income levels are different for the two suburbs, using all the data.

> (c) Test whether income levels are different separately for each age group.

12.4 Read in the apple tree yield data in [APPLE.DAT]. Test the independence of yield and spraying by analyzing the yield category versus the sprayed code. Is yield independent of spraying? Compute the ϕ measure of association.

NOTES

1. In some earlier versions of Minitab the "chi-square" (X^2) statistic reported by the TABLE command includes the continuity correction as described in Appendix 12A, and gives a value of 3.205 in this example.

13 COMPARISON OF SEVERAL POPULATIONS

INTRODUCTION

Minitab has several commands to perform one-way and balanced two-way analyses of variance, depending on the form of the data and how it is set up for analysis.

13.1 COMPARISON OF SEVERAL SAMPLES OF EQUAL SIZE

Data for several groups can be entered into Minitab with a separate column for each group. For instance, the data from Table 13.1 on reading scores of 18 children using three different workbooks has been typed into a file [WORKBOOK.DAT] with 6 lines, and three entries per line:

```
MTB > READ 'WORKBOOK.DAT' INTO C1-C3
      6 ROWS READ
  ROW   C1    C2   C3

    1    2     9    4
    2    4    10    5
    3    3    10    6
    4    4     7    3
      .    .    .

MTB > NAME C1 'BOOK1' C2 'BOOK2' C3 'BOOK3'
```

This format is often convenient for data with equal sample sizes in each group, but we must remember that the entries in any given row correspond to different observational units. For data in this form the command

AOVONEWAY on data in C,...,C

will perform a one-way analysis of variance as developed in the text. With our example data we have

```
MTB > AOVONEWAY ON 'BOOK1','BOOK2','BOOK3'

ANALYSIS OF VARIANCE
SOURCE      DF        SS        MS        F
FACTOR       2     84.00     42.00     22.50
ERROR       15     28.00      1.87
TOTAL       17    112.00
                                   INDIVIDUAL 95 PCT CI'S FOR MEAN
                                   BASED ON POOLED STDEV
LEVEL        N      MEAN     STDEV   ------+---------+---------+---------+
BOOK1        6     4.000     1.414   (-----*-----)
BOOK2        6     9.000     1.265                             (-----*-----)
BOOK3        6     5.000     1.414         (-----*-----)
                                   ------+---------+---------+---------+
POOLED STDEV = 1.366                  4.0       6.0       8.0      10.0
```

Minitab prints out the analysis of variance table, including the degrees of freedom (DF), and sum of squares (SS) for between groups (FACTOR), within groups (ERROR), and total, plus the within- and between-groups variances, or mean squares (MS), and their ratio (F). The F-ratio of 22.5 obtained here is compared with a value from an F-table for 2 and 15 degrees of freedom (such as Appendix X) to test whether there is significant between-group variation. Since this exceeds the value of $F_{2,15}(.01) = 6.3589$ the between-group variation is highly significant.

Minitab also prints out the number of observations, mean, and standard deviation in each group, and displays a plot with individual 95% confidence intervals for each group mean, based on the pooled within-group standard deviation (s_W).

13.2   COMPARISON OF SEVERAL SAMPLES OF UNEQUAL SIZE

The AOVONEWAY command does not require samples to be of equal size; it can analyze any one-way classification in the column format described above. However, it is frequently easier to work with data that have been coded with all of the measurements in a single column, one case per row, with a separate categorical variable to indicate the group membership. For instance, the data on TV viewing from Table 13.11 have been entered into a file with a code for the group membership (1-4) followed by the number of hours viewed per day. We read the data as follows:

```
MTB > READ 'TV.DAT' INTO C1 C2
      12 ROWS READ
  ROW    C1      C2

    1     1     5.0
    2     1     6.0
    3     1     4.0
```

```
   4     2     5.0
   .     .     .
```

```
MTB > NAME C1 'EDUCAT' C2 'TVHOURS'
```

The command

> **ONEWAY** AOV on C, levels in C

will perform a one-way analysis of variance on data in this form, as shown:

```
MTB > ONEWAY ON 'TVHOURS', LEVELS IN 'EDUCAT'

ANALYSIS OF VARIANCE ON TVHOURS
SOURCE     DF        SS        MS         F
EDUCAT      3    21.000     7.000      8.62
ERROR       8     6.500     0.813
TOTAL      11    27.500

                              INDIVIDUAL 95 PCT CI'S FOR MEAN
                              BASED ON POOLED STDEV
LEVEL       N      MEAN     STDEV   ---+---------+---------+---------+---
  1         3     5.000     1.000                          (-----*-----)
  2         4     4.000     0.816                    (----*----)
  3         3     3.000     1.000             (-----*-----)
  4         2     1.000     0.707   (------*------)
                                    ---+---------+---------+---------+---
POOLED STDEV = 0.901                0.0       2.0       4.0       6.0
```

The output formats from the AOVONEWAY and ONEWAY command are the same.

13.3 MULTIPLE COMPARISONS AND CONFIDENCE REGIONS

The summary statistics presented at the end of the ONEWAY output include everything that is needed to perform the multiple testing and simultaneous confidence procedures described in section 13.3 of the text. For instance, to get simultaneous confidence intervals on the differences in means in each pair of groups above, we create another Minitab data set with one line for each pair of groups, with the level, n, and mean for each group in the order $g, n_g, \mu_g, h, n_h, \mu_h$ as follows:

```
MTB > READ C5-C10
DATA>    1      3     5.000     2      4     4.000
DATA>    1      3     5.000     3      3     3.000
DATA>    1      3     5.000     4      2     1.000
DATA>    2      4     4.000     4      2     1.000
DATA>    2      4     4.000     3      3     3.000
DATA>    3      3     3.000     4      2     1.000
DATA> END OF DATA
```

```
      6 ROWS READ
MTB > NAME C5 'G' C6 'NG' C7 'MUG' C8 'H' C9 'NH' C10 'MUH'
MTB > NAME C11 'CI_HW' C12 'CI_L' C13 'CI_U'
MTB > LET 'CI_HW' = 0.901*3.355*SQRT(1/'NG'+1/'NH')
MTB > LET 'CI_L' = 'MUG' - 'MUH' - 'CI_HW'
MTB > LET 'CI_U' = 'MUG' - 'MUH' + 'CI_HW'
MTB > PRINT 'G' 'H' 'CI_L' 'CI_U'
  ROW    G    H     CI_L      CI_U

    1    1    2   -1.30874   3.30874
    2    1    3   -0.46815   4.46815
    3    1    4    1.24052   6.75948
    4    2    4    0.38213   5.61787
    5    2    3   -1.30874   3.30874
    6    3    4   -0.75948   4.75948
```

The pooled standard deviation is 0.901, and for an overall level of $\alpha = 0.06$ we have $\alpha^* = 0.01$ and $t_8(\alpha^*/2) = 3.355$. Then CI_{hw} is the half-width of the confidence interval, and CI_L and CI_U are the lower and upper confidence bounds on the difference in means for groups g and h. From these confidence intervals we can see that groups 1 and 2 are significantly different from group 4 (since the confidence intervals do not include 0), but no other comparisons are significant at an overall level of $\alpha = 0.06$.

13.4   A RANK TEST FOR COMPARISON OF SEVERAL DISTRIBUTIONS

The Kruskal-Wallis rank test for differences between groups is computed by the command

---
KRUSKAL-WALLIS test for data in C, levels in C

This command assumes the data are in the same format as for the ONE-WAY command in Section 13.2, and performs the assignment of ranks and the computation of the test statistic H. For instance, with the bottle cap machine data from Table 3.26 we have

```
MTB > READ C1 C2
DATA> 1 340
DATA> 1 345
DATA> 1 330
DATA> 1 342
DATA> 1 338
DATA> 2 339
DATA> 2 333
DATA> 2 344
DATA> 3 347
DATA> 3 343
DATA> 3 349
DATA> 3 355
```

```
DATA> END
     12 ROWS READ
MTB > NAME C1 'MACHINE' C2 'OUTPUT'
MTB > KRUSKAL-WALLIS TEST FOR 'OUTPUT' VERSUS GROUPS IN 'MACHINE'

LEVEL    NOBS    MEDIAN    AVE. RANK    Z VALUE
    1       5     340.0         4.8      -1.38
    2       3     339.0         4.7      -1.02
    3       4     348.0        10.0       2.38
OVERALL    12                   6.5

H = 5.656
* NOTE * ONE OR MORE SMALL SAMPLES
```

The H statistic is be compared with the value from a chi-squared table with $k - 1$ degrees of freedom to test the significance of the differences between groups. As described in the text, based on the chi-squared tables with 2 degrees of freedom the differences in the example are significant at the .10 level but not at the .05 level. The chi-square approximation is best for data with at least 5 observations per group, and Minitab warns us that we have less in some groups. Exact tables can be consulted for precise probability levels. If there are ties in the data, Minitab computes an adjusted statistic that takes account of the number of ties. It also would be referred to a chi-squared diftribution with $k - 1$ degrees of freedom to test group differences.

The z value for each group, computed as

$$z_g = \frac{\bar{r}_g - (n+1)/2}{\sqrt{(n+1)(n/n_g - 1)/12}},$$

where $\bar{r}_g = 1/n_g \sum_{i=1}^{n_g} r_{gi}$, is a measure of how the mean rank of the group differs from the overall mean rank. Under the null hypothesis, z_g is approximately normal with mean 0 and variance 1.

13.5 COMPARISONS FOR ONE CLASSIFICATION WHEN THE DATA ARE CLASSIFIED TWO WAYS

13.5.1 The F-Test

Data with two classifications of the groups are coded with two categorical variables to represent the group membership, plus a single score variable. For instance the beer can data from Table 13.16 was entered into a file [CAN.DAT] with the first field indicating the can (A = 1, B = 2, C = 3), the second field indicating the person, and the third field containing the "metallic" score:

```
MTB > READ 'CAN.DAT' INTO C1-C3
      15 ROWS READ
  ROW   C1   C2   C3

    1    1    1    6
    2    1    2    5
    3    1    3    6
    4    1    4    4
    .    .    .

MTB > NAME C1 'CAN' C2 'PERSON' C3 'SCORE'
```

The command

TWOWAY AOV for data in C versus levels in C and C

computes a two-way analysis of variance. This can be used for randomized block designs as well as other balanced designs with two factors. For the can data we obtain:

```
MTB > TWOWAY AOV FOR 'SCORE' VERSUS 'CAN' AND 'PERSON'

ANALYSIS OF VARIANCE  SCORE

SOURCE      DF        SS        MS
CAN          2    21.733    10.867
PERSON       4     9.333     2.333
ERROR        8     4.267     0.533
TOTAL       14    35.333
```

Here the treatments are cans, the blocks are persons, and the F-ratio for testing the differences between cans is the ratio of $MS(CAN)/MS(ERROR)$ = 20.4. This exceeds the .005 level of an F-table with 2 and 8 degrees of freedom, so the cans are highly significantly different.

13.5.3 Two-Way Classification with Repeated Observations

Two-way designs with multiple measurements per group can be analyzed with the TWOWAY command, if an equal number of measurements are available per group. For instance, the data from Table 13.32, with two assessments of the beer cans per person, are typed in a file [CAN2.DAT]. Each line contains the can number, person number, and assesment for a measurement: we have

```
MTB > READ 'CAN2.DAT' INTO C1-C3
      30 ROWS READ
  ROW   C1   C2   C3

    1    1    1    6
    2    1    1    6
```

```
3    1    2    4
4    1    2    6
.    .    .
```

```
MTB > NAME C1 'CAN' C2 'PERSON' C3 'SCORE'
MTB > TWOWAY AOV FOR 'SCORE' VERSUS 'CAN' AND 'PERSON'
```

```
ANALYSIS OF VARIANCE   SCORE
```

SOURCE	DF	SS	MS
CAN	2	43.47	21.73
PERSON	4	18.67	4.67
INTERACTION	8	8.53	1.07
ERROR	15	22.00	1.47
TOTAL	29	92.67	

The F-statistic for testing the differences between cans is $MS(CAN)/MS(ERROR) = 14.78$, which is compared with the entries in the F-table with 2 and 15 degrees of freedom. Again the cans are highly significantly different.

If we have reason to believe that the effects join in an additive fashion, and that there is no "interaction" between the two main effects, we can specify the

> *TWOWAY*
>
> ## ADDITIVE

subcommand to TWOWAY; this combines the interaction variation with the error variation. In our example this yeilds

```
MTB > TWOWAY AOV FOR 'SCORE' VERSUS 'CAN' AND 'PERSON';
SUBC> ADDITIVE.
```

```
ANALYSIS OF VARIANCE   SCORE
```

SOURCE	DF	SS	MS
CAN	2	43.47	21.73
PERSON	4	18.67	4.67
ERROR	23	30.53	1.33
TOTAL	29	92.67	

The TWOWAY command only prints out the analysis of variance table, and does not print mean levels. The TABLE command

> **TABLE** data with levels in C,...,C

will print statistics in each cell for an associated variable via the subcommand

> *TABLE*
>
> ## MEANS for variables C,...,C

For instance

```
MTB > TABLE FOR 'CAN' BY 'PERSON';
SUBC> MEANS OF 'SCORE'.

  ROWS: CAN    COLUMNS: PERSON

              1         2         3         4         5       ALL

     1   6.0000    5.0000    6.0000    4.0000    3.0000    4.8000
     2   2.0000    3.0000    2.0000    2.0000    1.0000    2.0000
     3   6.0000    4.0000    4.0000    4.0000    3.0000    4.2000
   ALL   4.6667    4.0000    4.0000    3.3333    2.3333    3.6667

   CELL CONTENTS --
          SCORE:MEAN
```

Other statistics are available via the subcommands N (the n of data points),
MEDIANS, SUMS, MINIMUMS, MAXIMUMS, STDEV, STATS (the same
as N plus MEANS plus STDEV), and DATA (a list of all data points).

COMMAND REFERENCE

Listed below are the general forms of the Minitab commands discussed in this chapter.

AOVONEWAY on data in C,...,C
ONEWAY AOV on C, levels in C
KRUSKAL-WALLIS test for data in C, levels in C
TWOWAY AOV for data in C versus levels in C and C
    ADDITIVE
TABLE data with levels in C,...,C
    N of values for variables C,...,C
    MEANS for variables C,...,C
    MEDIANS for variables C,...,C
    SUMS for variables C,...,C
    MINIMUMS for variables C,...,C
    MAXIMUMS for variables C,...,C
    STDEV for variables C,...,C
    STATS for variables C,...,C
    DATA for variables C,...,C

COMPUTER EXERCISES

13.1 Using the student grade data from Table 2.24 [STUDENT.DAT]:

(a) Test whether student major significantly affected the final grade, using both an F-ratio and a rank test.

(b) Test whether having logic training produced significantly different final grades using an F-ratio. Now do the same using a t-test. Do the results agree?

13.2 Set up a data set with the data from Table 13.31 on a picture ranking experiment. Are some paintings significantly more preferred than others?

13.3 Set up a data set with the data on weights of 12 tomatoes from Table 13.12. Obtain F-statistics for the effect of nitrogen, phosphate, and interaction on tomato weight. Locate the achieved levels of significance.

13.4 Examime effect of fertilization and irrigation on the yield of barley using the data in [BARLEY.DAT]. Test the significance of the effects assuming an additive model.

13.5 Examine the effects of sex and age on brain weight using data in [BRAIN.DAT]. Test the significance of the main effects and the interaction. Obtain a two-way table of cell means.

14 SIMPLE REGRESSION AND CORRELATION

INTRODUCTION

Minitab has extensive facilities for exploring the statistical relationship between two quantitative variables, including correlation and regression analysis.

14.2 STATISTICAL RELATIONSHIP

An important step in exploring the relationship between quantitative variables is graphing them. As we have seen previously, the command

PLOT data in C versus data in C

will result in a scatter plot of two variables. For instance, with the radioactive contamination and cancer mortality data from Table 14.3 we have

```
MTB > READ 'MORTAL.DAT' INTO C1 C2
       9 ROWS READ
   ROW     C1       C2

     1    8.34     210.3
     2    6.41     177.9
     3    3.41     129.9
     4    3.83     162.3
   .   .     .

MTB > NAME C1 'EXPOS' C2 'MORTAL'
```

When plotted, as shown on the next page, a clear relationship is apparent between the two variables, and the relationship seems roughly linear.

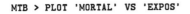

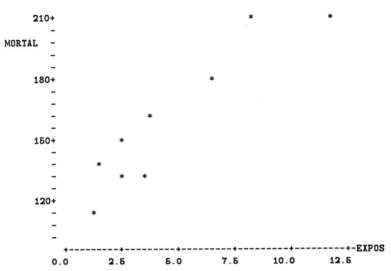

```
MTB > PLOT 'MORTAL' VS 'EXPOS'

       210+                              *            *
          -
MORTAL    -
          -
          -
       180+                         *
          -
          -
          -               *
          -
       150+          *
          -
          -     *
          -        *    *
          -
       120+
          -     *
          -
          -
          +---------+---------+---------+---------+---------+-EXPOS
        0.0       2.5       5.0       7.5      10.0      12.5
```

14.3 LEAST-SQUARES ESTIMATES

To fit a linear relationship between two variables by least squares
the command

> **REGRESS** data in **C** on **1** predictor in **C**

is used:

```
MTB > REGRESS 'MORTAL' ON 1 PREDICTOR IN 'EXPOS'

The regression equation is
MORTAL = 115 + 9.23 EXPOS

Predictor      Coef        Stdev      t-ratio
Constant     114.716       8.046       14.26
EXPOS          9.231       1.419        6.51

s = 14.01     R-sq = 85.8%    R-sq(adj) = 83.8%

Analysis of Variance

SOURCE       DF         SS          MS
Regression    1       8309.6      8309.6
Error         7       1373.9       196.3
Total         8       9683.5
```

The equation for the least-squares regression line is printed with several statistics about the regression, which we shall discuss below. The amount of output from the REGRESS command is controlled by the commands

BRIEF

and

NOBRIEF

The output above corresponds to the BRIEF option. If the command NO-BRIEF is given before a regression, then a complete list of the regression data is printed, including the independent value, the observed and predicted (fitted) dependent value, the (estimated) standard deviation of the predicted value, the residual (observed - predicted) value, and the standardized residual for each case:

Obs.	EXPOS	MORTAL	Fit	Stdev.Fit	Residual	St.Resid
1	8.3	210.30	191.71	7.05	18.59	1.54
2	6.4	177.90	173.89	5.32	4.01	0.31
3	3.4	129.90	146.19	4.97	-16.29	-1.24
4	3.8	162.30	150.07	4.80	12.23	0.93
5	2.6	130.10	138.44	5.50	-8.34	-0.65
6	11.6	207.50	222.17	11.00	-14.67	-1.69
7	1.3	113.50	126.25	6.68	-12.75	-1.04
8	2.5	147.10	137.70	5.56	9.40	0.73
9	1.6	137.50	129.67	6.32	7.83	0.63

The predicted and residual values can be saved in columns for further plotting or analysis with the extended command

REGRESS data in C versus **1** predictor in C, std. residuals in C, fitted values in C

For instance:

```
MTB > REGRESS 'MORTAL' ON 1 PRED IN 'EXPOS' PUT STD. RES. IN C3, FITS IN C4

[Output omitted]

MTB > NAME C3 'STDRESID' C4 'PRED'
```

The standardized (or Studentized) residuals are the residuals from the regression $y_i - \hat{y}_i$ divided by the (estimated) standard deviation of the residual $s_{y \cdot 12}$. We can plot both the observed and fitted values on the same plot with the command

MPLOT C vs C, C vs C, ...

as follows:

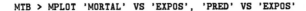

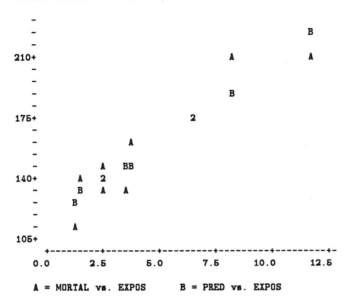

An "A" is plotted for the first pair of variables (observed data), and a "B" is shown for the second pair (fitted data). (The "2" indicates two points at the same plotting coordinate.) The fitted values lie on a line, which could be drawn by hand to produce a plot like Figure 14.5.

14.4 SAMPLING VARIABILITY; STATISTICAL INFERENCE

The output of the REGRESS command, as shown above, includes information on the total and residual sum of squares, as well as on the accuracy and significance of the regression coefficients. The analysis of variance table is similar to that discussed in Chapter 13, with the degrees of freedom, sum of squares, and mean square for regression, residual, and total (after subtracting the means of the variables). The F-ratio of MS(Regression) to MS(Residual) can be used to test the significance of the regression in the same way as in the analysis of variance.

The table of coefficients gives the fitted coefficients a and b of the regression line, the estimated standard deviations s_a and s_b of the coefficients, and the t-ratios a/s_a and b/s_b for testing the whether the coefficients are zero. The degrees of freedom for the t-ratios are the residual degrees of freedom from the regression (in this case, 7). Both the slope and intercept terms are highly significant. In the case of simple regression, the square of the t-ratio for the slope will equal the F-ratio for the regression discussed above.

The value of S = 14.01 is the residual standard deviation $s_{y \cdot x}$, the square root of MS(Residual).

14.5 THE CORRELATION COEFFICIENT: A MEASURE OF LINEAR RELATIONSHIP

The value of R^2 given in the REGRESS output is equal to the square of the correlation coefficient between the dependent and independent variable in the case of simple regression. ($R^2_{adj.}$, an adjusted R^2, will be discussed in the next chapter.) As described previously, the correlation coefficient r can be computed with the command

CORRELATION of C with C

For instance

```
MTB > CORRELATION OF 'EXPOS' AND 'MORTAL'

Correlation of EXPOS and MORTAL = 0.926
```

The significance of the correlation can be evaluated be comparing $t = \sqrt{n-2}(r/\sqrt{1-r^2})$ with values in the t-table with $n-2$ degrees of freedom.

14.6 RANK CORRELATION

The rank correlation can be computed by calculating the ranks of each variable with the command

RANKS of data in C, put in C

then applying the CORRELATION command, as follows:

```
MTB > RANKS OF 'EXPOS', PUT IN C5
MTB > RANKS OF 'MORTAL', PUT IN C6
MTB > CORRELATION OF C5 AND C6

Correlation of C5 and C6 = 0.833
```

The rank correlation is not quite as large the linear correlation in this case, but it is still large. The rank correlation will detect any monotone relationship between the variables, not just linear relationships.

COMMAND REFERENCE

Listed below are the general forms of the Minitab commands discussed in this chapter.

PLOT data in C versus data in C
REGRESS data in C versus **1** predictor in C, [std. residuals in C, [fitted values in C]]
BRIEF
NOBRIEF
MPLOT C vs C, C vs C, ...
CORRELATION of C with C
RANKS of data in C, put in C

COMPUTER EXERCISES

14.1 Using the data on heights and weights of 5-year old boys from Table 14.9 [BOY.DAT]:

(a) Compute the correlation of height and weight.

(b) Compute the rank correlation of height and weight.

(c) Regress weight on height. Estimate the mean and standard deviation of weights of 5-year olds who are 40" tall.

(d) Regress height on weight.

(e) Plot weight versus height, including both of the above fitted regression lines. Are the two lines the same?

14.2 Using the scores of 20 husbands and wives on a test of conformity from Table 14.17 [CONFORM.DAT]:

(a) Generate a scatter plot of the data. What does the plot indicate about the magnitude (high, medium, low) and direction of the correlation?

(b) Compute the correlation r.

14.3 Using the data on Republican percentage of the Presidential vote in Indiana counties in 1920 and 1952 [VOTE.DAT]:

(a) Find the least-squares regression of 1952 vote percentage on 1920 vote percentage.

(b) Compute a 99% confidence interval for the slope of the population regression line β.

(c) Test the hypothesis $H : \beta = 0$ against the alternative $\beta \neq 0$ at the 1% level.

14.4 Using the data on IQ's of identical twins from Table 14.20 [TWIN.-DAT] (see Exercise 14.22) compute the correlation r. By hand, compute the geneticity value $100r^2\%$ and the complimentary percentage $100(1 - r^2)\%$.

14.5 Compute the regression of achievement on aptitude for the data in Table 14.23 [APTITUDE.DAT]. Plot the fitted curve and actual data.

14.6 Using the auto noise pollution data from Exercise 5.22 (Table 5.67) [AUTO.DAT] compute the regression of y on x^2. Plot y versus x, including both the data and the fitted model.

14.7 Regress temperature on cricket chirps/minute using the data in [CRICKET.DAT] to obtain a prediction formula for temperature. Graph your formula and the data. What temperature would you guess it was if you heard one of these crickets chirping at 15 chirps/minute?

15 MULTIPLE REGRESSION AND PARTIAL CORRELATION

INTRODUCTION

The multiple regression abilities of Minitab are quite extensive and pass beyond the scope of the text. The ability to compute and test multiple regression models is among the most important uses of an interactive computer program.

15.1 REGRESSION ON TWO EXPLANATORY VARIABLES

15.1.1 The Least-Squares Estimates

The REGRESS command discussed in the preceeding chapter can be used for multiple regression as well in the form

> REGRESS data in C on K predictors in C,...,C

For instance, with the IQ data from Table 2.23 we can predict the initial reading score as a function of verbal and math IQ, as follows:

```
MTB > READ 'IQ.DAT' INTO C1-C5
     23 ROWS READ
  ROW    C1     C2     C3     C4     C5

    1     1     86     94    1.1    1.7
    2     2    104    103    1.5    1.7
    3     3     86     92    1.5    1.9
    4     4    105    100    2.0    2.0
    .     .     .

MTB > NAME C1 'PUPIL' C2 'VERBALIQ' C3 'MATHIQ' C4 'INITRD' C5 'FINALRD'
MTB > REGRESS 'INITRD' ON 2 PREDICTORS IN 'VERBALIQ' AND 'MATHIQ'

The regression equation is
INITRD = 0.285 + 0.0224 VERBALIQ - 0.00941 MATHIQ
```

```
Predictor        Coef        Stdev      t-ratio
Constant        0.2853      0.3355        0.85
VERBALIQ        0.022396    0.005135      4.36
MATHIQ         -0.009414    0.004773     -1.97

s = 0.1660      R-sq = 54.2%    R-sq(adj) = 49.6%

Analysis of Variance

SOURCE          DF          SS            MS
Regression      2         0.65300       0.32650
Error          20         0.55135       0.02757
Total          22         1.20435

SOURCE          DF        SEQ SS
VERBALIQ        1         0.54575
MATHIQ          1         0.10725
```

There are several subcommands to the REGRESS command, most of which are beyond our discussion. Useful subcommands include

REGRESS
> **RESIDUALS in C**

This puts the residuals $(y - \hat{y})$ in a column for further analysis.

REGRESS
> **NOCONSTANT**

This fits a regression model without the constant term.

REGRESS
> **COEFFICIENTS in C**

This puts the estimated coefficients in a column for further use (row 1 will contain the intercept, row 2 will contain the coefficient for the first predictor, etc.).

15.1.2   Statistical Inference

The multiple regression output from REGRESS is basically the same as for simple regression. The "Further Analysis of Variance" table divides the regression sum of squares among the dependent variables, when used in the order mentioned. The sum of squares for verbal IQ is for the simple regression of initial reading achievement. The sum of squares for math IQ is the additional regression sum of squares for both IQ scores together. If math IQ was listed first on the REGRESS command, the table would be presented with the sum of squares for math IQ alone, followed by the incremental sum of squares due to adding verbal IQ; of course, the total would be the same.

15.1.3 More Explanatory Variables

More explanatory variables can be included in the REGRESS command, as described above. The limit on the number of independent variables is not computational, but statistical. As the number of residual degrees of freedom decreases, the precision of the fitted regression coefficients decreases, and the regression model is affected more by purely random features of the data. Inclusion of irrelevant variables in the regression will increase the standard deviations of the estimated coefficients even for the relevant predictors; a proper choice of predictors is important. The Minitab STEPWISE command provides some tools for searching through several possible predictors to find the best model, but we shall not discuss it here.

15.1.4 Nonlinear Regression

For nonlinear relationships it is often possible to transform the model to a linear model by performing transformations on the independent or dependent variables or both, using the Minitab functions discussed previously. If the transformed model has a constant variance (not depending on the independent variables), then least-squares regression can be applied to the transformed model. A simple case is polynomial regression where the powers of the independent variable are used in a multiple regression model.

15.2 MULTIPLE CORRELATION

The multiple correlation $R_{y\cdot 12}$ is the square root of the R^2 reported by Minitab. R^2 is the proportion of the of sum of squares explained by the regression. Because the sample R^2 always increases when additional variables are added into the regression, Minitab also calculates an adjusted R^2 to take account of the overfitting of the data by multiple predictors; it is

$$R^2_{adj} = 1 - \frac{SS(\text{residual})/(n-p)}{SS(\text{total})/(n-1)}.$$

15.3 PARTIAL CORRELATION

The single correlations between several pairs of variables can be computed with the command

---
CORRELATION of data in C,...,C

For instance

```
MTB > CORRELATION OF 'VERBALIQ'  'MATHIQ' 'INITRD'

       VERBALIQ    MATHIQ
MATHIQ   0.769
INITRD   0.673     0.327
```

The correlation between math IQ and reading ability may be due mainly to the association between reading ability and verbal IQ and the association between verbal IQ and math IQ. The partial correlation of math IQ with reading ability after controlling for verbal IQ can be computed by computing the correlation of the residuals of each from regressions on verbal IQ, as follows:

```
MTB > NOTE PARTIAL CORRELATION
MTB > REGRESS 'INITRD' ON 1 PREDICTOR IN 'VERBALIQ';
SUBC> RESIDUALS IN C6.

The regression equation is
INITRD = 0.101 + 0.0146 VERBALIQ
[...]
MTB > REGRESS 'MATHIQ' ON 1 PREDICTOR IN 'VERBALIQ';
SUBC> RESIDUALS IN C7.

The regression equation is
MATHIQ = 19.6 + 0.827 VERBALIQ
[...]
MTB > CORRELATION OF C6 AND C7

Correlation of C6 and C7 = -0.404
```

The partial correlation between reading ability and math IQ after accounting for verbal IQ, is -0.404. The student may find it interesting to explain why higher math IQ is associated with lower reading ability, given verbal IQ.

COMMAND REFERENCE

Listed below are the general forms of the Minitab commands discussed in this chapter.

REGRESS data in C on **K** predictors in C,...,C
  RESIDUALS in C
  NOCONSTANT
  COEFFICIENTS in C
CORRELATION of data in C,...,C

COMPUTER EXERCISES

15.1 Read in the data from Table 15.3 on presidential elections [PRES-IDNT.DAT]. In the following, let y by standardized vote loss, and let x_1 and x_2 be the Gallup poll rating and the change in real income, respectively.

(a) Plot y, x_1, and x_2 versus year.

(b) Plot the pairs (x_1, x_2), (x_1, y), and (x_2, y).

(c) Calculate the multiple regression of y on x_1 and x_2. Test the significance of x_1 and x_2, and the overall significance of the regression.

15.2 Using the weather data in [WEATHER.DAT], regress January high temperatures on elevation and percent of sunshine. How much of the variance is accounted for? Compute 95% confidence intervals for the regression parameters. Do the signs of the coefficients agree with what you would expect? Do the same for the January low temperatures.

16 SAMPLING FROM POPULATIONS: SAMPLE SURVEYS

INTRODUCTION

Minitab can be used in a number of different ways to aid in generaing a random sample from a finite population. In this chapter we shall simply indicate some of the possibilities.

16.1 SIMPLE RANDOM SAMPLING

As seen previously, the command

```
SAMPLE K rows from C,...,C, put in C,...,C
```

draws a simple random sample without replacement of a specified size. The column on which the sample is drawn will usually consist of a unique identification number for each member of the sample frame, and the corresponding rows may contain information about the frame member. For instance, suppose we have a population of 1000 individuals, consisting of 450 males and 550 females identified by numbers 1 to 450 for males, and 451 to 1000 for females. We can create a Minitab version of the frame to sample from with the SET and CODE commands, as follows:

```
MTB > SET C1
DATA> 1:1000
DATA> END
MTB > CODE (1:450) TO 1 (451:1000) TO 2 IN C1, PUT IN C2
MTB > NAME C1 'ID' C2 'SEX'
MTB > NOTE SEX: MALE=1, FEMALE=2
```

We can then draw a simple random sample of size 10 from the frame and sort the results according to the ID numbers for ease of reference, as follows:

```
MTB > SAMPLE 10 VALUES FROM 'ID', CORR.VALUES FROM 'SEX', PUT IN C3,C4
MTB > NAME C3 'SAMP.ID' C4 'SAMP.SEX'
MTB > SORT 'SAMP.ID', CARRY 'SAMP.SEX', PUT IN 'SAMP.ID','SAMP.SEX'
MTB > PRINT 'SAMP.ID','SAMP.SEX'
 ROW  SAMP.ID  SAMP.SEX

   1     166       1
   2     182       1
   3     213       1
   4     214       1
   5     332       1
   6     414       1
   7     673       2
   8     674       2
   9     724       2
  10     878       2
```

Because we wanted to keep the correct sex with each identifier, during the sort, we used the command

---
SORT C, carry along C,...,C, put into C,...,C
---

that orders several columns according to the order in the first column, and maintains row relationships.

16.2   STRATIFIED RANDOM SAMPLING

If we have a categorical variable indicating the stratum membership in the frame, the COPY command

---
COPY C,...,C into C,...,C
---

with subcommands

COPY
   USE rows with C = K,...,K
---

and

COPY
   OMIT rows with C = K,...,K
---

can be used to extract the subframe for each stratum in turn, and SAMPLE can be used to obtain the appropriate size sample for the stratum. In the above example the simple random sample happened to choose 6 males and 4 females. If we wanted a stratified sample with 5 males and 5 females, we would proceed as follows:

```
MTB > COPY 'ID','SEX' PUT IN C5,C6;
SUBC> USE ROWS WITH 'SEX'=1.
MTB > NAME C5 'MALE.ID' C6 'MALE.SEX'
MTB > SAMPLE 5 ROWS FROM 'MALE.ID', CORRESP. OF 'MALE.SEX', PUT C7,C8
MTB > NAME C7 'M.S.ID' C8 'M.S.SEX'
MTB > COPY 'ID', 'SEX' , PUT IN C9,C10;
SUBC> USE ROWS WITH 'SEX' = 2.
MTB > NAME C9 'FEM.ID' C10 'FEM.SEX'
MTB > SAMPLE 5 ROWS FROM 'FEM.ID', CORRESP. OF 'FEM.SEX', PUT C11,C12
MTB > NAME C11 'F.S.ID' C12 'F.S.SEX'
MTB > STACK ('M.S.ID' 'M.S.SEX') TO ('F.S.ID' 'F.S.SEX'), PUT C13,C14
MTB > NAME C13 'SMP.ID' C14 'SMP.SEX'
MTB > SORT 'SMP.ID', CARRY ALONG 'SMP.SEX', PUT IN 'SMP.ID', 'SMP.SEX'
MTB > PRINT 'SMP.ID','SMP.SEX'
 ROW   SMP.ID   SMP.SEX
```

ROW	SMP.ID	SMP.SEX
1	14	1
2	143	1
3	155	1
4	303	1
5	318	1
6	618	2
7	655	2
8	663	2
9	855	2
10	904	2

Note that in analyzing the results from a stratified sample, if the stratum sizes are not proportional to the population proportions of the strata, the mean of all of the data is not an unbiased estimate of the population mean. Instead the means within strata should be computed and combined, weighted by the population proportions of the strata. Even when the sample sizes are proportional to strata sizes, the variances in the strata should be determined individually and combined by the formula in the text in order to estimate the standard deviation of overall sample mean.

16.3 CLUSTER SAMPLING

A cluster sample can be constructed by an iterated use of SAMPLE, first to choose the clusters, then to choose the sample within each cluster, in much the same fashion as demonstrated above.

16.4 SYSTEMATIC SAMPLING WITH A RANDOM START

A systematic sample with a random start can be obtained by dividing the population size N by the desired sample size n to obtain the sampling interval l (rounded to an integer). Then choose an integer at random from $1, \ldots, l$, say k, and select all cases of the form $k + (i - 1)l$, for

$i = 1, \ldots, n$. This can be done by finding all indentifiers I such that $I - k$ can be evenly divided by l; that is, all I such that $I = \lfloor (I - k)/l \rfloor \times l + k$, where $\lfloor x \rfloor$ is the greatest integer less than or equal to x. [This is computable in Minitab as $\text{ROUND}(x + .5) - 1$.] For example (K1 is N, K2 is n, K3 is l, and K4 is k):

```
MTB > NOTE K1: POPULATION SIZE, K2: SAMPLE SIZE, K3: SAMPLING INTERVAL
MTB > LET K1 = 1000
MTB > LET K2 = 10
MTB > SET C1
DATA> 1:K1
DATA> END
MTB > NAME C1 'ID'
MTB > LET K3 = ROUND(K1/K2)
MTB > RANDOM 1 DRAW, PUT IN C2;
SUBC> INTEGERS FROM 1 TO K3.
MTB > LET K4 = C2(1)
MTB > PRINT K3,K4
K3          100.000
K4          72.0000
MTB > LET C2 = 'ID' - (ROUND(('ID' - K4)/K3+.5)-1)*K3 - K4
MTB > NAME C2 'REM'
MTB > COPY 'ID' TO C3;
SUBC> USE ROWS WITH 'REM'=0.
MTB > NAME C3 'SMP.ID'
MTB > PRINT 'SMP.ID'
SMP.ID
     72     172     272     372     472     572     672     772     872     972
```

The cases in 'SMP.ID', printed above, identify the sample members.(Note the use of K1 in SETting the data into C1.)

16.5   SYSTEMATIC SUBSAMPLING WITH RANDOM STARTS

The above procedure can be applied several times with different starting values to obtain a systematic subsampling.

COMMAND REFERENCE

Listed below are the general forms of the Minitab commands discussed in this chapter.

SAMPLE K rows from C,...,C, put in C,...,C
SORT C, carry along C,...,C, put into C,...,C
COPY C,...,C into C,...,C
    USE rows with C = K,...,K
    OMIT rows with C = K,...,K

CODE (K,...,K) to K ... (K,...,K) to K for C,...,C store in C,...,C
STACK (E,...,E) on ... on (E,...,E) store in (C,...,C)

COMPUTER EXERCISES

16.1 Set up a frame for a population of 10 clusters of 100 individuals, each numbered from 1 to 100 within each cluster. Draw a cluster sample of 3 clusters, and 10 individuals per cluster.

16.2 Demonstrate how to draw 5 systematic subsamples of size 10 from a population of size 1000. [Hint: you may want to use SAMPLE to get the starting values, and use an EXEC file to perform the repeated sampling, then STACK the results together.]

DATA SETS

INTRODUCTION

Most of the data sets used in this *Guide* have been prepared as computer files for ease of access. The data sets are available from The Scientific Press on a $5\frac{1}{4}$ inch floppy disk for use with microcomputer versions of Minitab. This disk should be readable by all MS-DOS (IBM PC compatible) computers. In addition, the data sets have been provided to Minitab, Inc. for inclusion with their distribution tapes. It also should be fairly easy to transfer these data sets from the microcomputer diskette to any mainframe installation. Those not already having data transfer facilities might investigate the Kermit package, from Columbia, a public-domain terminal emulation and data transfer program available for many brands of computers. (For information, write to: KERMIT Distribution, Columbia University Center for Computing Activities, 7th Floor, Watson Laboratory, 612 West 115th Street, New York, N.Y. 10025.)

LIST OF DATASETS

For each data set, the variables are listed below. Data sets not described in the text also have a short description. Unless otherwise noted, the data are stored with one case per line, and can read with the READ command.

AIRLINE.DAT Airline Passengers per Month (Source: [3, p. 531])

1. International airline passengers: monthly totals (thousands)

Data for the period Jan. 1949 - Dec. 1960. (Stored horizontally, 12 cases per line; use SET to read.)

APPLE.DAT Apple Tree Yield (Source: [4])

1. "Sprayed" code (1 = sprayed, 2 = not sprayed)
2. Yield of apples (to tenths of a bushel)
3. Yield category (1 = 0 - 1.7 bushels, 2 = 1.8 - 2.5 bushels, 3 = over 2.5 bushels)

Fifteen apple trees were sprayed with a fungicide at the beginning of the summer and twenty trees were not sprayed. At the end of the summer, the apples were harvested and the yield of apples was tabulated for each tree.

APTITUDE.DAT Aptitude and Achievement Scores of 12 Students (Source: [1, Table 14.23])

1. Aptitude score
2. Achievement score

AREA.DAT Land Areas, by State (Source: [1, Table 2.20])

1. Region (1 = New England, 2 = Middle Atlantic, 3 = South Atlantic, 4 = East North Central, 5 = West North Central, 6 = East South Central, 7 = West South Central, 8 = Mountain, 9 = Pacific)
2. Total area (sq. mi.)
3. Farm area (1000 acres)

AUTO.DAT Average Speed and Noise Level in 30 Highway Sections (Source: [1, Table 5.67])

1. Section
2. Average speed (mph)
3. Noise level

BARLEY.DAT Barley Yield (Source: [5, p. 165])

1. Fertilization (1 = none, 2 = low, 3 = medium, 4 = high)
2. Irrigation (1 = no, 2 = yes)
3. Yield in bushels/acre

The results of an experiment on the effect of fertilization and irrigation on the yield of barley.

BEANS.DAT Bean Sorting (Source: [6, p. 213])

1. Worker ID (1-9)
2. Number of beans sorted by the first method
3. Number of beans sorted by the second method

In a study of the effectiveness of working methods, a firm wanted to compare two types of hand-picking methods (used in peanut and other industries). Nine participants in the test obtained the results in this data set (number of speckled beans hand-sorted from white and speckled beans within a given amount of time.

BOY.DAT Heights and Weights of 41 Five-Year-Old Boys (Source: [1, Table 14.9])

1. Boy
2. Height (inches)
3. Weight (pounds)

BRAIN.DAT Brain Weights (Source: [5, p. 167])

1. Brain weight (g)
2. Sex (1 = male, 2 = female)
3. Age (years) (1 = 50-80, 2 = 20-49)

Brain weights measured on 28 individuals.

BULBS.DAT Light Bulb Lifetimes (Source: [2, p. 3])

1. Light bulb lifetimes, in hours

Data on the lifetimes of 417 forty watt, 110 volt, internally frosted incandescant lamps taken from forced life tests.

CAMPER.DAT Individual Changes in Campers' Scores on Civil Liberties During the Summer (Source: [1, Table 9.20])

1. Score at beginning of summer
2. Score at end of summer

CAN.DAT Scores of Three Types of Can on "Metallic" Scale (Source: [1, Table 13.16])

1. Type of Can (A = 1, B = 2, C = 3)
2. Person
3. Score

CAN2.DAT Scores of Three Types of Can on Two "Metallic" Scale Replicates (Source: [1, Table 13.22])

1. Can (A = 1, B = 2, C = 3)
2. Person (1-6)
3. Score

CARBON.DAT Carbonate Environments Data (Source: [7, p. 363])

1. Environment (1 = clear,shallow water; 2 = abundant algae)
2. Eh below interface
3. pH below interface

4. Phi mean diameter

5. Phi standard deviation

Data describing the geochemical attributes of two different types of carbonate environments.

CONFORM.DAT Scores of 20 Husbands and Wives on a Test of Conformity (Source: [1, Table 14.17])

1. Couple

2. Husband's score

3. Wife's score

COOKIE.DAT Cookies and Cakes Ingredients (Source: assorted cookbooks and recipes)

1. Shortening (cups)

2. Sugar (cups)

3. Eggs (number)

4. Flour (cups)

5. Baking soda (teaspoons)

6. Baking powder (teaspoons)

7. Salt (teaspoons)

8. Other dry ingredients (cups)

9. Other liquid ingredients (cups)

10. Baking temperature (degrees Fahrenheit)

11. Baking time (minutes)

12. Type (cookie = 1, cake = 2)

The data are amounts of ingredients for various types of cookies and cakes.

CRICKET.DAT Cricket Chirping (Source: [8, p. 187])

1. Case number

2. Cricket chirps per minute

3. Temperature (degrees Fahrenheit)

Chirping rate of the striped ground cricket (Nemobius fasciatus fasciatus), at assorted temperatures.

DICHOTOMY.DAT Data for a Double Dichotomy (Source: [1, Table 5.1])

1. Name
2. Sex (1 = female, 2 = male)
3. University division (1 = graduate, 2 = undergraduate)

DRAFT.DAT 1969 Draft Lottery: Month of Birthday and Priority Number (Source: [1, Table 6.1])

1. Month of Birthday
2. Low priority (1-122)
3. Medium priority (123-244)
4. High priority (245-366)

FARM.DAT Data for 10 Farms (Source: [1, Table 3.20])

1. Farm
2. Number of children
3. Number of toys
4. Number of hogs
5. Number of chicks

FOOTBALL.DAT Weights and Heights of Stanford Football Players: 1970 (Source: [1, Table 2.16])

1. Weight (pounds)
2. Height (inches)

GASTAX.DAT State Gasoline Taxes (Source: [1, Table 2.19])

1. Tax rate (cents per gallon)

HAMBURGER.DAT Net Weights of 25 Packages of Hamburger Meat (Source: [1, Table 9.23])

1. Weight (pounds)

HEAD.DAT Head Measurements (Source: [1, Table 4.10])

1. Length (millimeters)
2. Breadth (millimeters)

HEART.DAT Heart Surgery Data (Source: [5, p. 335])

1. IQ at the beginning of the study
2. IQ at the end of the study
3. Treatment code (0 = nonsurgery, 1 = surgery)

In studying intelligence of children with heart disease, two groups of patients were considered: those who underwent surgery and those who did not. Investigations measured IQ at the beginning of the study and IQ at the end of the study.

HEIGHTS.DAT Heights of 11-Year-Old Boys (Source: [1, Table 2.11])

1. Midpoint
2. Frequency

INCOME.DAT Incomes of 49 Families (Source: [1, Table 2.17])

1. Respondent number
2. Income ($100's)

IQ.DAT IQ and Reading Achievement Scores of 23 Pupils (Source: [1, Table 2.23])

1. Pupil
2. Verbal IQ
3. Math IQ
4. Initial reading achievement
5. Final reading achievement

MORTAL.DAT Radioactive Contamination and Cancer Mortality (Source: [1, Table 14.3])

1. Index of exposure
2. Cancer mortality per 100,000 person-years

PRESIDNT.DAT Presidential Voting (Source: [1, Table 15.3])

1. Year
2. Party of current President (1 = Democrat, 2 = Republican)
3. Mean Congressional vote for party of current President in last 8 elections (%)

4. Nationwide Congressional vote for party of current President (%)

5. Standardized vote loss (%)

6. Gallup poll rating of President at time of election (%)

7. Current yearly change in real disposable income per capita ($)

READING.DAT Reading Scores of 30 Pupils Before and After Second Grade (Source: [1, Table 8.4])

1. Pupil

2. Reading score before second grade

3. Reading score after second grade

STATGRAD.DAT Final Grades in a Statistics Course (Source: [1, Table 10.4])

1. Student

2. Class (1 = undergraduate, 2 = graduate)

3. Final grade

STUDENT.DAT Data for 24 Students in a Statistics Course (Source: [1, Table 2.24])

1. Student Number

2. Major (1 = architecture (A), 2 = chemical engineering (CE), 3 = electrical engineering (EE), 4 = industrial administration (IA), 5 = materials sciences (M), 6 = psychology (P), 7 = public administration (PA), 8 = statistics (S))

3. Familiar with elementary logic (1 = yes, 0 = no)

4. Familiar with set theory (1 = yes, 0 = no)

5. Semesters of probability or statistics

6. Semesters of math analysis

7. Final grade

SWINE.DAT Weights of 58 Swine (Source: [1, Table 2.18])

1. Weight (pounds)

TENSILE.DAT Tensile Strengths (Source: [1, Table 9.22])

1. Strength (pounds per square inch)

TWIN.DAT IQ's of Monozygotic Twins Raised Apart (Source: [1, Table 14.20])

1. Case
2. Sex (male = 1, female = 2)
3. IQ of first born twin
4. IQ of second born twin

TV.DAT Average Number of Hours of TV Viewing per Day (Source: [1, Table 13.11])

1. Group (1 = not a high school graduate, 2 = high school graduate only, 3 = college graduate only, 4 = post-college study)
2. Average TV viewing (hours)

VOTE.DAT Republican Percentage of Total Presidential Vote in 1920 and 1952 for a Random Sample of 20 of the 92 Counties of Indiana (Source: [1, Table 14.20])

1. County
2. Percentage of vote in 1920
3. Percentage of vote in 1952

WEATHER.DAT Weather Data (Source: [10])

1. Census Division (1 = New England, 2 = Middle Atlantic, 3 = East North Central, 4 = West North Central, 5 = South Atlantic, 6 = East South Central, 7 = West South Central, 8 = Mountain, 9 = Pacific)

2. Elevation. Ground elevation of the city above sea level, in feet.

3. High temperatures for January. The average high temperature of the city for the month of January, based on the daily recorded highs for the month for the 30 year period 1931-1960. (degrees Fahrenheit)

4. Low temperatures for January. The average low temperature of the city for the month of January, based on the daily recorded lows for the month for the 30 year period 1931-1960. (degrees Fahrenheit)

5. Percent of sunshine. The average annual percent of sunshine during daylight hours, based on data recorded each day be the weather station. The period of years on which the data is based varies for each city and ranges from 6 to 95 years.

The National Weather Service collects data related to climate at hundreds of weather stations throughout the United States. This data set was obtained from historical weather data for 69 selected cities.

WORKBOOK.DAT Reading Achievement Scores of 18 Children Using Three Different Workbooks (Source: [1, Table 13.1])

1. Achievement score using workbook 1
2. Achievement score using workbook 2
3. Achievement score using workbook 3

LIST OF MINITAB COMMANDS

(Adapted from the *Minitab Handbook* [9], with permission.)

NOTATION

K denotes a constant, which can be either a number such as 8.3, or a stored constant such as K14.

C denotes a column, which must be typed with a C directly in front, such as C12. Columns may be named.

E denotes either a constant or a column.

M denotes a matrix, which must be typed with an M directly in front, such as M5.

[ ] denotes an optional argument.

Subcommands are shown indented, under the main command.

GENERAL INFORMATION

HELP explains Minitab commands
INFORMATION gives status of worksheet
STOP ends the current session

ENTERING NUMBERS

READ data [from **'FILENAME'**] into C,...,C
SET data [from **'FILENAME'**] into C
INSERT data [from **'FILENAME'**] between rows **K** and **K** of C,...,C
READ, SET, and INSERT all have the subcommands
    FORMAT (FORTRAN format)
    NOBS = K
The SET command allows patterns of the form $a : b$ [$/c$] for evenly spaced data. A stored constant may be used in such a pattern. A number before a list in parentheses repeats the entire list the indicated number of times. A number after a parenthesized list repeats each value individually the indicated number of times. This also is allowed with INSERT if operating on one column.

END of data
NAME for C is **'NAME'**, for C is **'NAME'** ... for C is **'NAME'**
RETRIEVE the worksheet saved [in **'FILENAME'**]

OUTPUTTING NUMBERS

PRINT the data in C,...,C
PRINT the data in K,...,K
WRITE [to **'FILENAME'**] the data in C,...,C
PRINT and WRITE have the subcommand
    FORMAT (FORTRAN format)

SAVE [in **'FILENAME'**] a copy of the worksheet

EDITING AND MANIPULATING DATA

DELETE rows K,...,K of C,...,C
INSERT data [from **'FILENAME'**] between rows K and K of C,...,C
    FORMAT (FORTRAN format)
    NOBS = K
COPY C,...,C into C,...,C
    USE rows K,...,K
    USE rows where C = K,...,K
    OMIT rows K,...,K
    OMIT rows where C = K,...,K
COPY C into K,...,K
COPY K,...,K into C
CODE (K,...,K) to K ... (K,...,K) to K for C,...,C store in C,...,C
STACK (E,...,E) on ... on (E,...,E) store in (C,...,C)
    SUBSCRIPTS into C
UNSTACK (C,...,C) into (E,...,E) ... (E,...,E)
    SUBSCRIPTS are in C
CONVERT ,using table in C,C, the data in C, put in C

You can refer to the missing value code * on a command or subcommand line by enclosing it in quotes, for instance:

```
CODE (-99) IN C1 TO '*', PUT BACK IN C1
```

You can use LET (see next section) to correct a number in the worksheet. For example:

```
LET C2(7) = 12.8
LET C3(5) = '*'
```

ARITHMETIC

ADD E to E ... to E, put into **E**
SUBTRACT E from E, put into **E**
MULTIPLY E by E ... by E, put into **E**
DIVIDE E by E, put into **E**
RAISE E to the power E, put into **E**
ABSOLUTE value of E, put into **E**
SQRT of E, put into **E**
LOGTEN of E, put into **E**
LOGE of E, put into **E**
ANTILOG of E, put into **E**
EXPONENTIATE E, put into **E**
ROUND to integer, put into **E**
SIN of E, put into **E**
COS of E, put into **E**
TAN of E, put into **E**
ASIN of E, put into **E**
ATAN of E, put into **E**
SIGNS of E, put into **E**
NSCORE normal scores of E, put into **E**
PARSUMS partial sums of E, put into **E**
PARPRODUCTS partial products of E, put into **E**

If the result of an arithmetic operation is undefined, the result is set to the * missing value code.

LET = expression
Expressions may use the arithmetic operators $+ - * /$ and $**$ (exponentiation) and any of the following: ABSOLUTE, SQRT, LOGTEN, LOGE, ANTILOG, EXPO, ROUND, SIN, COS, TAN, ASIN, ACOS, ATAN, SIGNS, NSCORE, PARSUMS, PARPRODUCTS, COUNT, N, NMISS, SUM, MEAN, STDEV, MEDIAN, MIN, MAX, SSQ, SORT, RANK, LAG. You can use subscripts to access individual numbers. For example:

```
LET C2 = SQRT(C1 - MIN(C1))
LET C3(5) = 4.5
```

COLUMN AND ROW OPERATIONS

COUNT the number of values in **C** [put into **K**]
N (number of nonmissing values) in **C** [put into **K**]
NMISS (number of missing values) in **C** [put into **K**]
SUM of the values in **C** [put into **K**]
MEAN of the values in **C** [put into **K**]
STDEV of the values in **C** [put into **K**]
MEDIAN of the values in **C** [put into **K**]
MINIMUM of the values in **C** [put into **K**]
MAXIMUM of the values in **C** [put into **K**]
SSQ (uncorrected sum of sq.) for **C** [put into **K**]

The following are all done rowwise.
RCOUNT of **C**,...,**C** put into **C**
RN of **C**,...,**C** put into **C**
RNMISS of **C**,...,**C** put into **C**
RSUM of **C**,...,**C** put into **C**
RMEAN of **C**,...,**C** put into **C**
RSTDEV of **C**,...,**C** put into **C**
RMEDIAN of **C**,...,**C** put into **C**
RMINIMUM of **C**,...,**C** put into **C**
RMAXIMUM of **C**,...,**C** put into **C**
RSSQ of **C**,...,**C** put into **C**

In computing column and row statistics, as with most other Minitab statistical procedures, entries with the * missing value code are excluded from the calculation.

PLOTS AND HISTOGRAMS

HISTOGRAM C,...,**C**
DOTPLOT C,...,**C**
Histogram and DOTPLOT both have the subcommands
    INCREMENT = K

START at **K** [end at **K**]
BY C
SAME scales for all columns

PLOT C versus C
MPLOT C versus C ... C versus C
LPLOT C versus C, using plotting symbols given by C
TPLOT C vs C vs C (three-dimensional plot)
PLOT, MPLOT, LPLOT, and TPLOT have the subcommands
    YINCREMENT = **K**
    YSTART at **K** [end at **K**]
    XINCREMENT = **K**
    XSTART at **K** [end at **K**]

TSPLOT [period = **K**] data in C
    INCREMENT = **K**
    START at **K** [end at **K**]
    ORIGIN = **K**
    TSTART at **K** [end at **K**]

WIDTH of all plots that follow is **K** spaces
HEIGHT of all plots that follow is **K** lines

BASIC STATISTICS

DESCRIBE C,...,C
    BY C
ZINTERVAL [with **K**% confidence] σ = **K** for C,...,C
ZTEST [of μ = **K**] σ = **K** on data in C,...,C
    ALTERNATIVE = **K**
TINTERVAL [with **K**% confidence] for C,...,C
TTEST [of μ = **K**] on data in C,...,C
    ALTERNATIVE = **K**
TWOSAMPLE test and c.i. [**K**% confidence] on C,C
    ALTERNATIVE = **K**
    POOLED procedure
TWOT test and c.i. [**K**% confidence] data in C, groups in C
    ALTERNATIVE = **K**
    POOLED procedure
CORRELATION between C,...,C [put into **M**]

COVARIANCE for C,...,C [put into **M**]
CENTER the data in C,...,C put into C,...,C
    LOCATION [subtracting **K**,...,**K**]
    SCALE [dividing by **K**,...,**K**]
    MINMAX [with **K** as min and **K** as max]

REGRESSION

REGRESS C on **K** predictors C,...,C [store standardized residuals in C
    [fits in C]]
    NOCONSTANT in equation
    WEIGHTS are in C
    MSE put into **K**
    COEFFICIENTS put into C
    XPXINV put into **M**
    HI put into C (leverage)
    RESIDUALS put into C (observed - fit)
    TRESID put into C (studentized, or deleted standardized residuals)
    COOKD put into C (Cook's distance)
    DFITS put into C
    VIF (variance inflation factors)
    DW (Durbin-Watson statistic)
    PURE (pure error lack-of-fit test)
    XLOF (experimental lack-of-fit test)
STEPWISE regression of C on the predictors C,...,C
    FENTER = **K** (default is 4)
    FREMOVE = **K** (default is 4)
    FORCE C,...,C
    ENTER C,...,C
    REMOVE C,...,C
    BEST K alternative predictors (default is 0)
    STEPS = **K** (default depends on output width)
NOCONSTANT in all stepwise and REFRESS that follow
CONSTANT return to fitting a constant in STEPWISE and REGRESS
BRIEF output [using print code = **K**] from REGRESS and STEPWISE
NOBRIEF return to default amount of output

ANALYSIS OF VARIANCE

AOVONEWAY analysis of variance for samples in C,...,C
ONEWAY data in C, subscripts in C [store residuals in C [fits in C]]
TWOWAY data in C, subscripts in C,C [store residuals in C [fits in C]]
    ADDITIVE model
    INDICATOR variables for subscripts in C, put into C,...,C

NONPARAMETRICS

RUNS test above and below **K** for data in C,...,C
STEST sign test [median = **K**] data in C,...,C
    ALTERNATIVE = **K**
SINT sign c.i. [**K**% confidence] data in C,...,C
WTEST Wilcoxon one-sample rank test [center = **K**] data in C,...,C
    ALTERNATIVE = **K**
WINT Wilcoxon c.i. [**K**% confidence] data in C,...,C
MANN-WHITNEY test and c.i. [alternative = **K**] [**K**% confidence] first
    sample in C, second sample in C
KRUSKAL-WALLIS test data in C, subscripts in C

TABLES

TALLY the data in C,...,C
    COUNTS
    PERCENTS
    CUMCNTS cumulative counts
    SUMPCTS cumulative percents
    ALL
CHISQUARE test on table stored in C,...,C
TABLE the data classified by C,...,C
    MEANS for C,...,C
    MEDIANS for C,...,C
    SUMS for C,...,C
    MINIMUMS for C,...,C
    MAXIMUMS for C,...,C
    STDEV for C,...,C
    STATS for C,...,C
    DATA for C,...,C

N for **C**,...,**C**
NMISS for **C**,...,**C**
PROPORTION of cases = **K** [thru **K**] in **C**,...,**C**
COUNTS
ROWPERCENTS
COLPERCENTS
TOTPERCENTS
CHISQUARE analysis [output code = **K**]
MISSING level [for each classification variable **C**,...,**C**]
NOALL in margins
ALL for **C**,...,**C**
FREQUENCIES are in **C**
LAYOUT **K** by **K**

TIME SERIES

ACF [with up to **K** lags] for series in **C** [put into **C**]
PACF [with up to **K** lags] for series in **C** [put into **C**]
CCF [with up to **K** lags] between series in **C** and **C**
DIFFERENCES [of lag **K**] for data in **C**, put into **C**
LAG [by **K**] data in **C**, put into **C**
ARIMA p = **K**, d = **K**, q = **K**, data in **C** [put residuals in **C** [predicteds in **C** [coefficients in **C**]]]
ARIMA p = **K**, d = **K**, q = **K**, P = **K**, D = **K**, Q = **K**, S = **K**, data in **C** [put residuals in **C** [predicteds in **C** [put coeff in **C**]]]
    CONSTANT term in model
    NOCONSTANT term in model
    STARTING values are in **C**
    FORECAST [forecast origin = **K**] up to **K** leads ahead [store forecasts in **C** [confidence limits in **C**,**C**]
BRIEF output [using print code = **K**] for ARIMA and REGRESS

See also TSPLOT above.

EXPLORATORY DATA ANALYSIS

STEM-AND-LEAF display of **C**,...,**C**
    INCREMENT = **K**

    TRIM "outliers"
    BY C
    SAME increment on all displays
BOXPLOTS for C [levels in C]
    LINES = K
    NOTCHES
    LEVELS K,...,K
MPOLISH C, levels in C,C [put residuals in C [fits in C]]
    COLUMNS (start iteration with column medians)
    ITERATIONS = K
    EFFECTS put common into K, rows into C, columns into C
    COMPARISON values, put into C
RLINE y in C, x in C [put residuals into C [fits into C [coeff. into C]]]
    MAXITER = K (maximum number of iterations)
RSMOOTH C, put rough into C, smooth into C
    SMOOTH by 3RSSH, twice
CPLOT (condensed plot) y in C versus x in C
    LINES = K (plot length)
    CHARACTERS = K (plot width)
    XBOUNDS are from K to K
    YBOUNDS are from K to K
CTABLE (coded table) data in C, row levels in C, column levels in C
    MINIMUM value in each cell should be coded
    MAXIMUM value in each cell should be coded
    EXTREME value in each cell should be coded
ROOTOGRAM data in C [bin boundaries in C]
    BOUNDARIES store bin boundaries in C
    DRRS store double root residuals in C
    FITTED values, store them in C
    COUNTS store them in C
    FREQUENCIES are in C [bin boundaries are in C]
    MEAN = K
    STDEV = K
LVALUES of C [put letter values in C [mids in C [spreads in C]]]

SORTING

SORT the values in C carry along corresponding rows of C,...,C put into
    C put corresponding rows into C,...,C
RANK the values in C put ranks into C

DISTRIBUTIONS AND RANDOM DATA

RANDOM K observations into each of **C,...,C**
PDF for values in **E** [store results in **E**]
CDF for values in **E** [store results in **E**]
INVCDF for values in **E** [store results in **E**]
RANDOM, PDF, CDF, and INVCDF have the subcommands
    BINOMIAL $n = $ **K**, $p = $ **K**
    BERNOULLI trials with $p = $ **K** (RANDOM only)
    POISSON mean $= $ **K** (0 ¡ mean ¡ 101)
    INTEGERS uniform on $a = $ **K** to $b = $ **K**
    DISCRETE distribution with x values in **C**, probabilities in **C**
    NORMAL [$\mu = $ **K**, $\sigma = $ **K**] (default $\mu = 0$, $\sigma = 1$)
    UNIFORM continuous [on $a = $ **K** to $b = $ **K**] (default $a = 0$ $b = 1$)
    T degrees of freedom $= $ **K**
    F df numerator $= $ **K**, denominator $= $ **K**
    CAUCHY [$a = $ **K**, $b = $ **K**] (default $a = 0$, $b = 1$)
    LAPLACE [$a = $ **K**, $b = $ **K**] (default $a = 0$, $b = 1$)
    LOGISTIC [$a = $ **K**, $b = $ **K**] (default $a = 0$, $b = 1$)
    LOGNORMAL [$\mu = $ **K**, $\sigma = $ **K**] (default $\mu = 0$, $\sigma = 1$)
    CHISQUARE with d.f $= $ **K**
    EXPONENTIAL [$b = $ **K**] (default $b = 1$)
    GAMMA $a = $ **K**, $b = $ **K**
    WEIBULL $a = $ **K**, $b = $ **K**
    BETA $a = $ **K**, $b = $ **K**

SAMPLE K rows from **C,...,C** put into **C,...,C**
BASE for random number generator $= $ **K**

MATRICES

READ data [from **'FILENAME'**] into a **K** by **K** matrix **M**
PRINT M,...,M
COPY C,...,C into **M**
COPY M into **C,...,C**
COPY M into **M**
TRANSPOSE M into **M**
INVERT M into **M**
DIAGONAL is **C**, form into **M**

DIAGONAL of **M**, put into **C**
EIGEN for **M** put values into **C** [vectors into **M**]

In the following commands **E** can be **C**, **K**, or **M**
ADD E to **E**, put into **E**
SUBTRACT E from **E**, put into **E**
MULTIPLY E by **E**, put into **E**

MISCELLANEOUS

NOTE comments may be put here
ERASE E,...,**E**
RESTART begin fresh Minitab session
NEWPAGE start next output on a new page
OW output width = **K** spaces
OH output height = **K** lines
PAPER (put terminal output on paper)
NOPAPER
OUTFILE 'FILENAME' (put output in this file)
NOOUTFILE (put output just to the terminal)
BATCH batch mode
TSHARE interactive or timesharing mode

The symbol # anywhere on a line tells Minitab to ignore everything after that on a line.

To continue a command onto another line, end the first line with the symbol &.

STORED COMMANDS AND LOOPS

The commands STORE and EXECUTE provide both a simple macro (or stored command file) capability and a simple looping capability.
STORE [in **'FILENAME'**] following commands
   (Minitab commands go here)
END of stored commands
EXECUTE commands [in **'FILENAME'**] [**K** times]
NOECHO the commands that follow
ECHO the commands that follow

The integer part of a column number may be replaced by a stored constant. This is useful in loops. For example:

```
LET K1 = 5
PRINT C1 - CK1
```

Since K1 = 5, this PRINTS C1 through C5.

REFERENCES

[1] Anderson, T. W. and Stanley L. Sclove (1986). *The Statistical Analysis of Data*. 2nd ed. The Scientific Press, Palo Alto.

[2] Bowker, Albert H. and Gerald J. Lieberman (1972). *Engineering Statistics*. 1st ed. Prentice Hall, Englewood Cliffs, New Jersey.

[3] Box, George E. P. and Gwilym M. Jenkins (1976). *Time Series Analysis*. Rev. ed. Holden Day, Inc., San Francisco.

[4] Conover, William J. (1971). *Practical Nonparametric Statistics*. John Wiley and Sons, New York.

[5] Dunn Olive J. and Virginia A. Clark (1974). *Applied Statistics: Analysis of Variance and Regression*. John Wiley and Sons, New York.

[6] Kreyszig, Erwin (1970). *Introductory Mathematical Statistical Principles and Methods*. John Wiley and Sons, New York.

[7] Krumbein, W. C. and Franklin A. Graybill (1965). *An Introduction to Statistical Models in Geology*. McGraw-Hill, New York.

[8] Larsen, Richard J. and Donna F. Stroup (1976). *Statistics in the Real World: A Book of Examples*. Macmillan, New York.

[9] Ryan, Barbara F., Brian L. Joiner, and Thomas A. Ryan, Jr. (1985). *Minitab Handbook*. 2nd ed. Duxbury Press, Boston.

[10] U. S. Bureau of the Census (1973). *Statistical Abstract of the United States*. U.S. Department of Commerce, Washington.

INDEX